J. L. R. Proops · M. Faber · G. Wagenhals

In Cooperation with
S. Speck · G. Müller · F. Jöst · P. Gay

Reducing CO_2 Emissions

A Comparative Input-Output-Study
for Germany and the UK

With 74 Figures and 56 Tables

Springer-Verlag
Berlin Heidelberg New York
London Paris Tokyo
Hong Kong Barcelona
Budapest

Dr. John L. R. Proops
Keele University, Department of Economics
Keele, Staffordshire ST5 5BG, United Kingdom

Professor Dr. Malte Faber
Alfred-Weber-Institut für Sozial- und Staatswissenschaften
der Universität Heidelberg, Grabengasse 14
D-6900 Heidelberg, FRG

Professor Dr. Gerhard Wagenhals
Institut für Volkswirtschaftslehre 520
Universität Hohenheim, Postfach 70 05 65
D-7000 Stuttgart 70, FRG

ISBN 3-540-55947-7 Springer-Verlag Berlin Heidelberg New York Tokyo
ISBN 0-387-55947-7 Springer-Verlag New York Berlin Heidelberg Tokyo

This work is subject to copyright. All rights are reserved, whether the whole or part of the material is concerned, specifically the rights of translation, reprinting, reuse of illustration, recitation, broadcasting, reproduction on microfilms or in other ways, and storage in data banks. Duplication of this publication or parts thereof is only permitted under the provisions of the German Copyright Law of September 9, 1965, in its version of June 24, 1985, and a copyright fee must always be paid. Violations fall under the prosecution act of the German Copyright Law.

© Springer-Verlag Berlin · Heidelberg 1993
Printed in Germany

The use of registered names, trademarks, etc. in this publication does not imply, even in the absence of a specific statement, that such names are exempt from the relevant protective laws and regulations and therefore free for general use.

2142/7130-543210 – Printed on acid-free paper

Preface

The global greenhouse effect may be one of the greatest challenges ever to face humankind. If fossil fuel use, and the consequent CO_2 emissions, continue to increase at their current trend, there is the possibility that over the next century there will be massive climate change and the flooding of coastal areas.

The economics profession is beginning to respond to this challenge, through seeking to understand the economic processes which determine the demand for energy, the proportion of this energy supplied by fossil fuels, and the policy instruments available for reducing fossil fuel demand while still supplying appropriate amounts of energy.

This study is a contribution to that literature. We examine the impact of structural changes in the German and UK economies upon CO_2 emissions over the last two decades, and explore the potential for further structural change to reduce such emissions. This study is different from much of the current literature, in that we do not presuppose that the respective economies consist of only one, or a few, sectors. Instead, we analyse the interrelationships of 47 sectors for about 20 years, using input-output methods. We also deal with the effects of the changing sectoral structure of imports and exports of these two countries on the 'responsibility' for CO_2 emissions. On the basis of this extensive evidence we have a solid foundation to develop different scenarios to show how the 'Toronto target' of reducing CO_2 emissions by 20% over 20 years can be achieved.

One of the many interesting results is that this reduction can be achieved while maintaining a 2% growth rate of GDP and high employment, without taking recourse to more nuclear power. Another outcome is that the structure and the developments of CO_2 emissions in both countries are surprisingly similar. This gives rise to the hope that our findings are much more general than we originally thought. Thus they could, in particular, be employed to outline a corresponding emission reduction policy for Europe, and to some extent for the developed world at large.

The major conclusion of our study is that, if we have the will to reduce CO_2 emissions, then this can be done without major economic disruption. The prerequisite is that this will is maintained for a sustained period. There can be no 'quick fixes' to the global greenhouse effect.

This book is addressed to economists, environmental analysts, statisticians and policy makers. However, it is not intended to be simply a research monograph. We have been at pains to present the material in such a way that it can be used also as a textbook of input-output methods applied to environmental analysis. For this reason we have extensively discussed our conceptual framework for ecological economic modelling and paid close attention to the effects of structural changes at both global and national levels.

This research was mainly funded by the Commission of the European Community (Stimulation Programme for Economic Science) under research grant SPES-0013-UK(A). We are grateful for this considerable financial support.

We thank Reiner Manstetten and Armin Schmutzler for their constructive comments.

We also thank, for their patience and constructive assistance, the following bodies: Deutsches Statistisches Bundesamt, the UK Central Statistical Office, Eurostat and the UN Statistical Office.

We now indicate the authors principally responsible for drafting each chapter, and making the relevant computations, by their initials.

Ch. 1: MF, JP, FJ. Ch. 2: MF, GW. Ch. 3: FJ, GM, MF. Ch.4: JP, SS, GM, FJ. Ch. 5: JP, SS, GM, GW. Ch. 6: MF, GW. Ch.7: JP. Ch. 8: JP. Ch. 9: JP, SS, GM. Ch. 10: JP, SS, GW. Ch. 11: GW. Ch. 12: JP, PG. Ch. 13: JP, SS. Ch. 14: JP, MF.

JOHN PROOPS   Keele
MALTE FABER   Heidelberg
GERHARD WAGENHALS   Stuttgart, Hohenheim

Stefan Speck   Keele
Georg Müller   Heidelberg
Frank Jöst   Heidelberg
Philip Gay   Keele

July 1992

Contents

Part I. Introduction 1

1 Introduction 3

  1.1 The Study 3
  1.2 The Greenhouse Effect and Global Warming 3
    1.2.1 The Natural Greenhouse Effect 3
    1.2.2 The Anthropogenic Greenhouse Effect 5
    1.2.3 The Evidence for Global Warming 6
  1.3 Industrialisation and the Control of Nature 7
  1.4 The Difficulty of Political Implementation of CO_2 Targets ... 11
    1.4.1 Wish and Will: The Need for Consensus 12
    1.4.2 Reductions in CO_2 Emissions: Local, National and Global Levels 13
  1.5 Practical Implementation of CO_2 Reduction 15
  1.6 The Contents of the Book 17

2 Concepts and Methods for the Study 19

  2.1 Introduction 19
  2.2 Our Point of Departure 19
  2.3 The CO_2 Problem 20
  2.4 General Principles for Ecological Economic Modelling 21
    2.4.1 Empirical Foundations 22
    2.4.2 Time 23
    2.4.3 Natural Boundary Conditions 23
    2.4.4 Non-Linear Relationships 24
    2.4.5 Production 24
    2.4.6 Demand 25
    2.4.7 Quantities and Prices 25
    2.4.8 Modelling Ex Post and Ex Ante 26
  2.5 Our Aims 26
  2.6 Our Method 27

Part II. Understanding CO_2 Emissions ... 31

3 The Problem of Climate Change and CO_2 Emissions ... 33

- 3.1 Introduction ... 33
- 3.2 The Nature of the Problem ... 33
- 3.3 The Evidence ... 34
  - 3.3.1 Global Greenhouse Gas Emissions ... 34
- 3.4 Historical Trends ... 36
  - 3.4.1 Conversion Units ... 36
  - 3.4.2 Data on Energy Use, CO_2 Emissions and GDP ... 37
- 3.5 Review of CO_2 Modelling in the Economics Literature ... 40
  - 3.5.1 Factors Determining CO_2 emissions ... 41
    - 3.5.1.1 Output and Population Growth ... 42
    - 3.5.1.2 Energy Efficiency ... 42
    - 3.5.1.3 Energy Price ... 42
    - 3.5.1.4 Back-Stop Technologies ... 43
  - 3.5.2 Key Determinants of the Costs of Abating GHG Emissions ... 43
  - 3.5.3 Emission Reduction Scenarios ... 44
    - 3.5.3.1 Reduction Targets ... 44
    - 3.5.3.2 Policy Instruments ... 45
    - 3.5.3.3 Growth Effects ... 45
- 3.6 Outlook and Conclusions ... 45

4 Decomposing the Rate of Change of CO_2 Emissions ... 47

- 4.1 Introduction ... 47
- 4.2 Understanding the Time Structure of CO_2 Emissions ... 47
  - 4.2.1 The Influential Variables ... 48
- 4.3 CO_2 Emissions as a Product of Several Variables ... 48
- 4.4 Decomposing the Components of CO_2 Emission Changes ... 49
  - 4.4.1 Necessary Extensions to the Technique ... 51
- 4.5 The Rates of Change of CO_2 Emissions ... 51
- 4.6 Conclusions ... 54

A4 Appendix: Decomposition with Differencing ... 55

- A4.1 Introduction ... 55
- A4.2 The Differential Approach to Decomposition ... 55
- A4.3 Approximating the Differential with the Difference ... 57
- A4.4 Resolving Ambiguity: Forward, Backward and Central Differences ... 57

A4.5 Calculating the Differencing Remainder Term	59
A4.6 Differencing Error-Prone Empirical Data	61

5 CO_2 Emissions by Germany and the UK — 64

5.1 Introduction	64
5.2 Energy Use, CO_2 Emission and GDP for Germany and the UK	64
5.2.1 Data Sources	65
5.2.2 The CO_2/Energy Ratio and the Fuel Mix	66
5.3 Energy Use and GDP	67
5.4 Energy Use and CO_2 Emission by Fuel Types	70
5.5 Energy Use and CO_2 Emission by Households and in Production	71
5.6 Production CO_2 Emission by Sector	72
5.7 Decomposition of German and UK CO_2 Emission Changes	74
5.7.1 Decomposition of Total CO_2 Emissions Changes	74
5.7.2 Decomposition of Production CO_2 Emissions Changes	76
5.7.3 Decomposition of Households CO_2 Emissions Changes	77
5.7.4 Decomposition of Total CO_2 Emissions Changes	78
5.8 Sectoral Decomposition of CO_2 Emissions Changes	79
5.8.1 Decomposition of Overall Rates of Change of CO_2 Emission by Sectors	82
5.8.2 Decomposition of Aggregate CO_2 Emission Changes	83
5.8.3 A Full Sectoral Decomposition of UK Production CO_2 Emissions Changes	86
5.9 Conclusions	87

Part III. Modelling Approach — 89

6 A Framework for Modelling Production — 91

6.1 Introduction	91
6.2 Invention, Innovation and Irreversibility	91
6.3 Activity Analysis vs. Production Functions	92
6.4 Activity Analysis and Input-Output Analysis	95
6.4.1 Input-Output Analysis and National Accounting	96
6.4.2 Activity Analysis and Dynamics	97
6.5 Input-Output Approaches versus General Equilibrium Models	97
6.6 Conclusions	98

7 Input-Output Methods 99

7.1 Introduction 99
7.2 Inter-Industry Trading and Input-Output Tables 99
    7.2.1 A Two Sector Input-Output Model 100
    7.2.2 Activity Analysis as the Production Assumption 101
    7.2.3 Output Structure in Equation Form 103
7.3 Matrix Representation 104
    7.3.1 The Unit Matrix and Matrix Inversion 105
    7.3.2 Solving Simultaneous Equations in Matrix Form 106
7.4 The Input-Output Model in Matrix Form 107
    7.4.1 Reconstructing the Input-Output Table 109
7.5 Decomposing Direct and Indirect Effects 111
7.6 Prices in Input-Output Models 113
    7.6.1 Values and Prices in Input-Output Tables 114
7.7 Constructing Input-Output Tables from Value-Based Data 116
    7.7.1 Deflating Input-Output Coefficients 118
7.8 Input-Output Tables and National Accounts 119

8 The Analysis of CO_2 Emissions with Input-Output Methods 121

8.1 Introduction 121
8.2 The Input-Output Assessment of CO_2 Emissions 121
    8.2.1 An Input-Output Model of Production CO_2 Emissions 122
    8.2.2 An Input-Output Model of Final Demand CO_2 Emissions 124
    8.2.3 Production CO_2 Emissions from Non-Fossil Fuel Sources 125
    8.2.4 An Equation for Total CO_2 Emissions 127
8.3 Comparing CO_2 Emissions Over Time and Between Countries 128
8.4 Imports, Exports and 'Attributable' CO_2 Emissions 131
    8.4.1 Exports and the Attribution of CO_2 Emissions 131
    8.4.2 Imports and the Attribution of CO_2 Emissions 132
    8.4.3 Calculating CO_2 Emission in a 2-Region, n-Sector Economy 133
8.5 The Sensitivity of CO_2 Emissions to Changed Parameters 138
    8.5.1 The Elasticities of CO_2 Emission with the Parameters 139
    8.5.2 The Derivative of CO_2 Emission with Respect to a_{ij} 140
    8.5.3 The Derivative of CO_2 Emissions with Respect to c_{if} 142
    8.5.4 The Derivative of CO_2 Emissions with Respect to p_{if} 143
    8.5.5 The Derivative of CO_2 Emissions with Respect to y_i 143
    8.5.6 The Elasticities of CO_2 Emissions 144

Contents xi

    8.5.7 The Elasticities of CO_2 Emissions with Respect to a_j and a_i 144
    8.6 Conclusions 146

Part IV. Data Analysis 147

9 German and UK Input-Output Data for Studying CO_2 Emissions 149

    9.1 Introduction 149
    9.2 The Data Requirements 149
    9.3 Data Collection and Processing 150
        9.3.1 The Aggregation of the Input-Output Tables 151
    9.4 Data Sources Used 152
        9.4.1 Production of the **A** Matrices for Germany 152
        9.4.2 Production of the **A** Matrices for the UK 152
        9.4.3 Production of the **C** Matrices for Germany 153
        9.4.4 Production of the **C** Matrices for the UK 153
        9.4.5 Production of the **P** Matrices for Germany 154
        9.4.6 Production of the **P** Matrices for the UK 155
        9.4.7 Production of the **Z** Matrices for Germany 155
        9.4.8 Production of the **Z** Matrices for the UK 155
        9.4.9 Production of the **e** Vector for Germany 155
        9.4.10 Production of the **e** Vector for the UK 156
        9.4.11 Production of the **m** Vectors for Germany 156
        9.4.12 Production of the **m** Vectors for the UK 156
        9.4.13 Production of the **B** Matrices for Germany 156
        9.4.14 Production of the **B** Matrices for the UK 156
        9.4.15 Production of the **u** Vectors and Y for Germany 156
        9.4.16 Production of the **u** Vectors and Y for the UK 157
        9.4.17 Further Adjustment of the UK Data 157
    9.5 The Structure of CO_2 Emission: Germany 1988 157
        9.5.1 The Nature of Input-Output Data 157
        9.5.2 The Basic Data Used in the Analysis 158
        9.5.3 The CO_2 Intensities 158
        9.5.4 Attributed CO_2 Emissions 161
        9.5.5 CO_2 Intensities and Emissions by Sector 164
        9.5.6 CO_2 Emission Elasticities 164
    9.6 Conclusions 169

10 Input-Output Analysis of German and UK CO_2 Emissions 170

10.1 Introduction 170
10.2 Attribution of CO_2 Emissions by Germany and the UK 170
10.3 Imports and Exports: CO_2 Emission and Responsibility ... 174
10.4 Changes in CO_2 Emission over Time 177
   10.4.1 The Full Sectoral Decomposition 179
   10.4.2 The Aggregate Decomposition 183
10.5 Differences in CO_2 Emissions between Germany and the UK 188
10.6 Conclusions 192

A10 Appendix: Decomposition of Changes in CO_2 Emissions 193

A10.1 Introduction 193
A10.2 Tables 193

Part V. Scenarios 203

11 Scenario Simulations 205

11.1 Introduction 205
11.2 Changing the Structure of Final Demand 206
   11.2.1 Final Demand for Non-Fuel Goods 206
   11.2.2 Final Demand for Fuels 207
   11.2.3 From Private to Public Transport 208
   11.2.4 Simulation Results 209
   11.2.5 Conclusion 210
11.3 Changing the Energy Efficiency 210
   11.3.1 Efficiency of Industrial Fuel Use 211
      11.3.1.1. Electricity Generation 211
      11.3.1.2 Iron and Steel Industry 212
      11.3.1.3 Building Materials Industry 212
      11.3.1.4 Food Processing Industry 213
      11.3.1.5 Simulation Results 213
   11.3.2 Efficiency of Direct Final Demand Fuel Use 214
      11.3.2.1 Housing Insulation 214
      11.3.2.2 Household Appliances 215
      11.3.2.3 District Heating 215
      11.3.2.4 Simulation Results 215
   11.3.3 Conclusion 216
11.4 Changing the Fuel Mix 216
   11.4.1 Natural Gas 217

11.4.2 Non-fossil Fuels	218
11.4.2.1 Nuclear Energy	219
11.4.2.2 Conclusion	220
11.4.2.3 Renewable Energy Sources	220
11.4.3 Conclusion	222
11.4.4 Simulation Results	222
11.5 Trend Extrapolations	223
11.6 A Sequence of Plausible Scenarios	224
11.7 Conclusions	226

12 A 'Minimum Disruption' Approach to Scenario Analysis 227

12.1 Introduction	227
12.2 The 'Minimum Disruption' Approach	227
12.3 Minimising Disruption of Final Demand	229
12.3.1 The Constraint on CO_2 Emission Reduction	229
12.3.2 Minimising y Disruption Subject to CO_2 Emission Target	230
12.3.3 Subject to CO_2 Emission Target and GDP Growth Target	232
12.3.4 Subject to CO_2 Emission Target and Employment Target	235
12.3.5 Subject to CO_2 Emission, GDP and Employment Targets	238
12.4 Minimising Disruption of Fuel Use Coefficients	240
12.4.1 Minimising Disruption of C, with Constant Energy Efficiency	241
12.5 Minimising Disruption of Inter-Industry Trading	244
12.6 Conclusions	246

13 'Minimum Disruption' Scenario Simulations 247

13.1 Introduction	247
13.2 Data Sources	247
13.3 Changes in Final Demand	248
13.3.1 Changes in Final Demand with no Other Constraints	248
13.3.2 Demand Changes with a GDP Growth Constraint	250
13.3.3 Demand Changes with an Employment Growth Constraint	252
13.3.4 Demand Changes with GDP and Employment Growth Constraints	252
13.4 Changes in Fuel Mix and Fuel Efficiency	255
13.4.1 Change in Fuel Mix with no Constraints	257

        13.4.2 Change in Fuel Mix with Constant Energy Efficiency Constraint 257
   13.5 Changes in the Structure of Inter-Industry Trading 264
   13.6 Conclusions 267

Part VI. Policy 269

14 Policy Conclusions for Reducing CO_2 Emissions 271

   14.1 Introduction 271
   14.2 Major Conclusions 271
        14.2.1 History 271
        14.2.2 Analysis 272
        14.2.3 Scenarios 274
   14.3 Policy Overview 275
   14.4 The Need for the *Will* 276

References 278

Author Index 286

Subject Index 289

List of Figures 295

List of Tables 298

Part I

Introduction

1 Introduction

1.1 The Study

In this study we explore how economic structural change has brought about increased atmospheric concentrations of carbon dioxide (CO_2), and how economic structural change may be used to reduce CO_2 emissions over the next 20 years. The principal analytical technique we use is input-output analysis. This has the benefit that an economy can be studied in terms of many interacting productive sectors.

In Section 1.2 we briefly review the natural and anthropogenic 'Greenhouse Effects', and the impact of likely global warming. Then, in Section 1.3, we seek to set the global greenhouse effect in its context, by examining how industrialisation is intimately related to the control of nature. Section 1.4 discusses the difficulty in achieving the political implementation of emission targets, but notes that the economic structural change that may be required to achieve such targets may be comparable to that achieved under other circumstances, such as war. To achieve such targets, there is the need not only to 'wish' that they be achieved, but there is also the need for the 'will' to achieve them. In Section 1.5 the problems of the practical implementation of CO_2 reduction are discussed. Finally, Section 1.6 outlines the contents of the rest of this book.

1.2 The Greenhouse Effect and Global Warming

This study is prompted by the growing concern of the likely environmental impact of global warming, because of the increasing emission of CO_2 and other 'Greenhouse' gases. In this section we briefly outline the nature of this problem.

1.2.1 The Natural Greenhouse Effect

The average surface temperature of the Earth is about 15°C; the temperature of interstellar space is around −250°C. The difference between these two temperatures can be attributed to two sources; the radiant energy of the sun, and the nature of the Earth's atmosphere.

The Earth's atmosphere consists mostly of nitrogen and oxygen, with greater or lesser traces of other gases, including water vapour and carbon dioxide (CO_2). Both water vapour and CO_2 have the property that while they are transparent to visible light, they absorb infrared radiation[1] relatively strongly. Now the wavelength of the radiation emitted by a body depends upon that body's temperature; the higher the temperature, the shorter the wavelength. Thus the sun, with a surface temperature of several thousand °C, emits its radiation mainly in the relatively short wavelengths of visible radiation. (It is visible because we have evolved eyes suited to the dominant radiation wavelengths.) The Earth, when heated by the sun's rays, also emits radiation, but as the temperature of the Earth's surface is much lower than that of the sun's, the wavelength of the emitted radiation is much longer. The great bulk of the Earth's radiation is at infrared wavelengths.

We can now understand the role of water vapour and CO_2 in the Earth's temperature maintenance; radiation from the sun at visible wavelengths passes freely through the atmosphere, and a proportion is absorbed by the Earth's surface. The warmed Earth's surface in turn radiates, but in the infrared, and this radiation is quite strongly absorbed by the water vapour and CO_2 in the atmosphere. This absorbed energy is then re-emitted, again in the infrared, but whereas the initial radiation was all being radiated *away* from the Earth, now some of the radiation from the atmosphere is back *towards* the Earth. The consequence is that the total amount of radiation striking the Earth is increased, so the average temperature of the Earth's surface is increased.

This transparency of water vapour and CO_2 to visible light, but relative opaqueness to infrared radiation, is also a property of glass. Hence a greenhouse (or glasshouse) is very efficient at retaining incident solar radiation[2]. Because of this similarity in effect, the warming of the Earth by water vapour and CO_2 is now known as the 'Greenhouse Effect'. As this effect is 'natural' (i.e. involves no human intervention), it is better called the 'Natural Greenhouse Effect'.

---

[1] Readers without a background in natural science may be alarmed at the 'radiation' reaching the Earth from the sun. Here the term is used in its scientific sense, of something that is radiated, rather than its popular, journalistic sense, of something that comes from nuclear reactors and nuclear weapons. The radiation discussed here is properly termed 'electromagnetic radiation'. Of this type of radiation, only the very short wave X-rays and gamma rays are of the latter, harmful, type.

[2] However, a considerable proportion of the effectiveness of greenhouses can be attributed to their reducing the cooling flow of air over their contents.

Now if the Earth's atmosphere contained no water vapour or CO_2, then it has been calculated the average surface temperature would be $-19°C$. Clearly, without the natural greenhouse effect it would be impossible to sustain present forms of life over most of the Earth's surface. In terms of the health and well-being of the human race, the natural greenhouse effect is a *very* good thing!

1.2.2 The Anthropogenic Greenhouse Effect

Economic activity requires the use of 'free energy'; i.e. energy which is available for doing work. Early societies used almost entirely organic materials for this purpose, either as foodstuffs (e.g. grains, meat) or fuels (e.g. wood), where the free energy they embody derives almost directly from sunlight. However, the rise of industrialisation both required and, through the accumulation of capital goods, allowed, the use of much more intense energy sources than those derivable directly from sunlight. Fortunately for the industrialisation process, the Earth is well endowed with 'fossil fuels', in the form of coal, oil and gas[3]. Over the past three hundred years these have been exploited more and more intensively, to provide fuels for the production and use of new capital goods.

All fossil fuels are carbon based, and their combustion to release free energy generates CO_2. This release of CO_2 into the atmosphere has been extensively documented, and it is estimated that the concentration of atmospheric CO_2 has risen by one-third since 1800. This rise accounts for approximately two-thirds of the estimated release of CO_2 from burning fossil fuels. It assumed that the remaining CO_2 has been absorbed into the oceans.

The effect of this increase in atmospheric CO_2 concentration is to increase the greenhouse effect. As this contribution to the overall greenhouse effect is human in source, we term it the 'Anthropogenic Greenhouse Effect'. As discussed in more detail in Chapter 3, agricultural and industrial activity generates other greenhouse gases, including methane, nitrous oxide and chlorofluorocarbons (CFCs). The overall effect of these trace gases will be to increase the surface temperature of the Earth. The actual size of the temperature rise to be expected, and the precise dynamics of how it

---

[3] There is general agreement that coal is fossilised organic material from earlier times. However, there is some debate about whether naturally occurring oil and gas are also derived from early life, or whether they derive from free interstellar methane, captured at the time of the Earth's coalescence.

occurs, is far from easy to estimate. This is because the Earth's climate is mediated by two very complex systems, the atmosphere and the oceans, which themselves interact in a complex way. However, it is not our aim to attempt to contribute to this assessment, as we are not climatologists. However, among climatologists a consensus seems to be emerging on at least the effect of doubling atmospheric CO_2 concentration, and this is that the global average temperature would rise by 2-5°C.

1.2.3 The Evidence for Global Warming

There is now an almost unanimous agreement among climatologists that the emissions of greenhouse gases contribute in an essential way to the change of the global climate. Further, this climate change will have far-reaching consequences for all life on Earth. As discussed above, the main cause is the trace gas carbon dioxide (CO_2), mainly produced because of the burning of coal, oil and gas (the fossil fuels), and the non-renewable use of biomass, to supply available energy. The results from climate research show that the contribution of CO_2 to the anthropogenic greenhouse effect is 55% of the total [IPCC 1990].

Since the industrial revolution, the nature of economic activity in the West has been that of increasing control over the rest of nature, largely through the powering of capital equipment with fossil fuels. The world use of fossil fuels is rising rapidly, largely through the industrialisation of the Third World, though energy demand in the USA and the EC is still also increasing, at a modest rate (see Chapter 3 for details). For the foreseeable future, there is every likelihood that global energy use, and therefore fossil fuel use and CO_2 emissions, will continue to rise.

These circumstances have led to a suggestion from the Toronto Conference in 1988. There it was proposed that the major industrialised countries take as a target for CO_2 emissions a 20% reduction from the 1988 emissions by 2005. This is equivalent to an annual reduction in CO_2 emissions of somewhat more than 1.3% per annum.[4]

However this target has not been widely adopted. Instead, EC countries have generally taken as their target the *stabilisation* of CO_2 emissions at 1990 levels by 2000. In our view this is a lamentably feeble response, and one which it is trivial to achieve. We therefore take the '20% reduction

---

[4] This target was actually suggested from the floor of the meeting, during general discussion [Pearce 1991a].

over 20 years' target as the one we assess in this study. For simplicity, we take a 1% per annum reduction of CO_2 emissions, over twenty years, as our version of the 'Toronto Target'.

1.3 Industrialisation and the Control of Nature

Having outlined the problem of global warming, we wish to set this problem in its context. To this end, in this section we discuss the relationship between industrialisation and the non-human natural world.

In contrast to an agricultural based economy, an industrial economy draws much more on the resources of the natural environment and its repercussions on nature are much greater. We can distinguish seven stages in modern Western economic activity.

i Resources such as coal, iron, copper and minerals are extracted from nature.

ii These are used to build up stocks of capital goods, such as tools, machines, buildings, cars, trains, ships, aircraft, roads, railways, harbours, airports, and last but not least, weapons.

iii In turn, these capital goods are used to provide consumption goods and services, which require the use of available energy, derived from naturally occurring fuels.

iv Because of the Second Law of Thermodynamics [cf. Georgescu-Roegen 1971], the above three types of activity all involve the emission of waste gases, waste liquids and waste solids. According to the First Law of Thermodynamics, which has been operationalised in economics as the 'Materials Balance Approach', all material and energy which enters the production process has, at some time, to leave it. Hence, all inputs into economic activity will at some time be an output into the natural environment.

Because of the requirement of materials balance, we see that the greater is the input of materials into the productive system, the greater must be the eventual output of material to the natural environment.

v The emissions to nature will be partly degraded, and partly utilised by natural systems. However, if the quantity of emissions becomes very great, and its toxicity increases, the capacity of the natural environment to cope with these emissions may not be sufficient, and these emissions may accumulate as stocks.

vi If this process of waste emissions is continued, there are two effects. The capacity of the natural environment to degrade anthropogenic emissions may be impaired, or even destroyed. For example, high nitrate and phosphate levels in lakes, from fertiliser run-off and detergent use, lead to rapid growth in algae populations. This blocks off sunlight, reducing photosynthesis by water based plants, with a consequent fall in oxygen concentrations, and decline in the populations of fish and aquatic insects. This process is known as 'eutrophication'. At the stage when anthropogenic waste begins to impair natural environmental activities, we refer to it as 'pollution'. At the present the *flows* of emissions are in the forefront of the analysis.

vii Here it is useful to note that, from an anthropocentric perspective, nature can be considered as a vast, and in many cases only partly known and understood, set of 'natural capital goods' [cf. Faber, Manstetten and Proops 1992a, 1992b].

Because of the overloading, impairment or destruction of the natural degradation capacity, there will be growing stocks of pollution. These growing stocks of noxious materials will further impair natural environmental activities. However, as most natural systems are strongly non-linear in their relations, so as to maintain robustness and stability, their response to growing pollution stocks is also often non-linear. There may be a 'critical level' of pollution, below which the system seems to cope, but above which there is a dramatic change in the qualitative characteristics of the system [cf. Faber and Proops 1990:Ch. 5].

The length of time between when anthropogenic impact on the natural system begins, and when the natural system exhibits catastrophic change, will depend on the scale of the impact on the system, and on the scale of the system itself. Large systems exhibit great inertia, and it may, for example, be several months or years between the time a lake is overloaded with pollution, and the time it is noticed that the lake is 'dying'. In the terminology of flows and stocks, which is akin to capital theory, this means that it takes time for the flows of pollution to build up to stocks which reach critical levels.

Regarding CO_2 emissions, as discussed above, anthropogenic CO_2 emissions have been increasing for two hundred years, as has atmospheric CO_2 concentration. So far, the global climate system has not exhibited catastrophic change. However, also as discussed above, it may do so within the next century, if present CO_2 emission trends continue.

Considering these seven stages, we note that there has been an increasing awareness of the environmental effects of modern economic activity during the past three decades. While up to the beginning of the seventies the availability of non-renewable resources was not of particular concern, this changed dramatically with the first oil price shock. During the past two decades this awareness has been maintained, and perhaps even sharpened. This occurrence was perhaps the first time in modern history that the natural environment, in its role of a supplier of resources, was felt to be limiting to economic activity, not only regionally, but also globally.

The environment has been used as a sink for waste industrial products for hundreds of years. Up until the mid-seventies this was perceived to be a source of only local environmental degradation, mostly of a transient type. With the rise of the environmental movement, and the European Green parties, this perception has altered radically, with action being taken to reduce the flows of pollutants into the environment. Examples of the success of these actions are the significant improvements in the water quality of Lake Erie and the Rhine. The rise in the awareness of these environmental problems was mirrored in the economics profession by the rise of *Environmental Economics* as a discipline, taking as its major analytical tools the theory of externalities and public goods. The focus of environmental policy was upon the polluters and those who suffered from the flows of pollution, but paid little attention to the effects of pollution on the natural environment; i.e. the building up of stocks of noxious materials. The conceptual framework of this analysis and policy formulation was taken to be that common to most economics; viz. rational individuals maximising their well-being, subject to constraints that were perfectly known, or at worst known stochastically.

While the attention of the environmental issue was mainly on the flows of pollutants into the natural environment, in recent years attention has shifted to the problem of the accumulation of stocks of pollution, on a global scale [c.f. Dasgupta 1983]. This has two consequences for economic analysis and policy making. First, it is no longer sufficient to consider only the relationships between the polluters and the *human* sufferers from pollution. On a global scale, the effects of pollution are principally upon global ecosystem health, and only indirectly, through this, upon human welfare [Costanza et al. 1992]. For example, if increased atmospheric CO_2 concentrations cause the global climate to alter, the direct effect of this climate change may be modest and seemingly benign. Few dwellers of Northern Europe would complain about warmer summers. However, the potential disruption to agricultural production, water supply and disease spread may be extreme, and very disbeneficial to most humans. To differentiate these kinds of problems from those environmental problems which are caused by the flow of pollutants we call them 'ecological problems'. We note in passing that with the becoming of awareness there was at the same time the rise of a new discipline: *Ecological Economics* [cf. Costanza 1991].

In contrast to the perceived scarcity of natural resources, or the perceived effects of flows of pollution, the consequences of increasing stocks of pollutants are difficult to assess, and one does not know when these consequences will occur. How can one behave as a rational economic agent in the face of such radical uncertainty?

An economist might approach this question in the following way. The problem can be set up as one of cost-benefit analysis. She would assume that marginal benefits will be positive, but decreasing, with pollution abatement activity, while the marginal costs of this activity will be also positive, but increasing. The optimal degree of pollution abatement would be where marginal costs equal marginal benefits. Thus a deterministic optimal outcome would be sought with this approach.

We do not consider this type of cost-benefit approach to be adequate to assessing the problem of global environmental change, for three reasons.

1. We face radical uncertainty about the nature of the outcome of such change.

2. We are also uncertain as to when this change will be encountered.

3. Even if we were in a position to know possible outcomes, at least stochastically, the normal decision making approach to such risky outcomes would be inappropriate, as the 'experiment' is one we are unlikely to be able to repeat indefinitely, so it is difficult to interpret the mathematical expectation, which we would form by such a technique.

Instead of this optimising, cost-benefit approach, we think that in such circumstances it is appropriate to adopt a precautionary approach. This would entail the establishment of limits and targets to govern the relationship between economic activity and the natural world. Following Daly's [1991] notion of 'ecological Plimsoll lines'[5], we would use the best scientific information to assess the level at which stocks of pollutants are

---

[5] A 'Plimsoll line' is a line marked on the hull of a ship, to indicate the greatest depth to which it may be safely laden.

likely to become damaging to the global environment, and to ensure that this stock of pollutant does not exceed, or is reduced to, no more than this Plimsoll target.[6]

In the case of pollutants which are still being generated, a modification of this simple stock target, which can be adopted as an interim measure, is the flow target. For example, one might aim to stabilise the emission of a pollutant by a target date, and perhaps thereafter seek to reduce the emission to a lower target by a later date. This is precisely the procedure adopted regarding CFC emissions, under the Montreal Protocol, and what has been suggested for CO_2 emissions at the Toronto conference.

In the light of the above discussion of the nature of global ecological problems, of which the anthropogenic greenhouse effect is one, and the widespread use of the flow target approach to pollution control, in the rest of the study we exclusively use CO_2 emission targets for both the assessment of past CO_2 emission, and of scenarios of future CO_2 emission. We never take the usual cost-benefit approach to the greenhouse effect.

1.4 The Difficulty of Political Implementation of CO_2 Targets[7]

To achieve target CO_2 emission reductions requires that two conditions be met. First, that the targets be feasible in the required timescale, subject to whatever social constraints are imposed. Second, that it is possible to achieve a will to achieve these targets, as reflected in a consensus within the various nation states concerned, and between those nation states. The first condition, the feasibility, is dealt with at length in the remainder of this study. For the moment we concentrate upon the second condition, achieving the will and the consensus to attain the target.

---

[6] In contrast to the usual cost-benefit approaches, referred to above, where the damage is rising but finite with the environmental load, in the case of the use of the Plimsoll line, the value damage function is infinite as soon as the Plimsoll line is reached [cf. Dasgupta 1983:49ff].

[7] The argument developed in this section are based on our experience with the American and German water legislation and the German waste legislation (see Brown and Johnson [1982], Faber and Stephan [1987], Faber, Stephan and Michaelis [1989]).

1.4.1 Wish and Will: The Need for Consensus

In political reality there seems to exist a paradox concerning major environmental issues. Many people (e.g. 86% of a representative sample in Germany in a poll in 1990) consider that the environmental question is the most important one. Nevertheless, no *really* drastic policy change has been carried through.

To examine this seemingly paradoxical social behaviour, it is pertinent to distinguish between the '*wish*' and the '*will*'. The *wish* is only the desire to have a better environment and to be less exposed to possible ecological catastrophes in the future. In contrast, the *will* contains not only the desire for such a change, but also the determination to carry it through, with all its consequences. In particular, this may require the maintenance of the effort for the appropriate length of time, even if that is very long. The main difference between the wish and the will is that the latter implies one is prepared to take into account, and to endure, all the frictions, the strenuous efforts and sufferings which the wish entails.

To give an example, many Americans in 1941 had the wish to contribute to the ending of the Nazi Regime under Hitler in Germany. But only after the Japanese attack on Pearl Harbour had taken place were the majority of the people of the United States willing to undertake all the necessary actions; the building up of a war-time economy, the recruiting, the fighting and also the dying, which was to occur.

It is our contention that major environmental changes would similarly lead to structural change, of production methods in particular, and of economic behaviour in general. Of course, this would not occur in such a drastic way and in such a short time period of only four years, as for the American experience in World War II, but instead more gradually and over a much longer time period (i.e. over several decades).

The main economic reasons why such major change in environmental policies leads to frictions and, in many instances, are painful, is that it causes structural change. This in turn leads to at least income and wealth redistribution, because of rising unemployment and decreasing profits in some sectors of the economy, and vice versa in others. It is evident that on short-run or medium-run considerations, there will be at least a strong and very active minority, if not even a majority, which is against any drastic changes leading to such negative consequences concerning their economic welfare.

However, this part of the population may also be convinced to agree to such changes if longer time-horizons are taken into account. Only then is it possible to transform the *wish* for better environmental conditions, into a *will*, which leads to corresponding actions over long periods.

We consider one means to achieve this aim politically is to change the constitution (be it written, as in Germany, or unwritten as in the UK). This can only be done in general if there is a wide *consensus*, of at least two-thirds of the population. This consensus seems to us to be a prerequisite for the realisation of such ecological targets as to reduce CO_2 emissions by 20% of their 1985 quantities by 2005 (see Faber and Manstetten [1989], Petersen [1992], Faber, Stephan and Michaelis [1989]).

We are convinced, from our historical findings in Chapters 3 and 5, and our scenario analyses in Chapters 11 and 13, that there exists enough flexibility in market economies to allow them to change themselves.

Finally we note two further insights from the American war experience. One concerns an example of what the *will* is able to achieve. Shortly after Pearl Harbour, Roosevelt called in his industrial advisers and announced that the war would need to be fought in both the European and Pacific theatres and that he wanted "50,000 planes". His advisers were at pains to point out that it would take several years to be able to build a fleet of such size, given the very modest US air forces at that time. "Oh no", Roosevelt replied, "I meant 50,000 planes *per year*!". By the end of the war, just four years later, the US economy was producing 70,000 planes per year.

The second refers to a general insight. The structural change of the American economy during World War II, each year brought about a considerable improvement of economic welfare for all US citizens. We believe that if the industrialised countries really face up to the ecological question, this will unleash enormous inventive and productive potential. We should not be surprised if this led also to increased economic welfare at large. We do not consider that the cost of this restructuring for ecological ends will be growth foregone. On the contrary, this restructuring, with its massive scrapping of old capital and generation of new, more efficient capital, will lead to more rapid rates of growth after a relatively short initial adjustment period. This assertion is even more the case if national income is properly measured, to take account of the portion of national income generated by the natural environment [Faber and Proops 1991a].

1.4.2 Reductions in CO_2 Emissions: Local, National and Global Levels

The greenhouse effect is a global problem, since the CO_2 emissions are into the global atmosphere, and have global effects. Hence a reduction on a local or national level has an extremely small effect on the total amount of CO_2 emissions, and almost no impact for the well-being of those who made the reductions, even in the long-run. Therefore, we are confronted with a repeated 'Prisoners' Dilemma' situation [Axelrod 1984]; in particular we face the problem of the 'Free Rider'. This always occurs if a public good has to be supplied. It is supposed to be 'rational' to let others carry the efforts and to enjoy the benefits without incurring any costs.

So one might ask: "What use is it if Germany or the UK reduce their CO_2 emissions, while other countries do not?" The answer to this question is not easy. However, for its solution we consider the following aspects to be pertinent.

i   The industrialised countries have been responsible for about 80% of the CO_2 emissions in the past. This means that they have made use of too great a part of the Earth's common capacity to absorb CO_2. Hence, the industrialised countries have, from a point of justice and fairness, good reasons to be first in reducing CO_2 emissions.

ii The industrialised countries have the highest incomes. It is well known that 'environmental quality' is one of those goods which are demanded only after the elementary wants have been satisfied. Therefore one cannot expect all countries in the world to start at the same time to reduce CO_2 emissions.

iii The industrial countries have not only the means, in terms of income and wealth, but also the know-how to invent and innovate appropriate techniques, as well as the know-how and institutions to implement corresponding legislation, be it in terms of emission limits or via taxes or permits.

iv Finally, it is worth noting that those who start first, although they carry the considerable start-up costs, later have great advantages. This is so because they can export the corresponding techniques, and also they have made the structural adjustments earlier. A telling example of the latter case is a statement in the *Financial Times* that the UK chemical companies complain that they are no longer competitive with their German rivals. The reason they give is the following: in Germany the adjustment to stringent water purity legislation already took place in the 70s and early 80s, so the German companies have already adjusted their operations, techniques and capital stock, via investment, while the UK companies now face this investment as UK legislation comes into force.

1.5 Practical Implementation of CO_2 Reduction

To implement reduction targets for CO_2 emissions in practice, there are two groups of instruments which are extensively discussed in environmental policy:

i Regulative laws; e.g. emission limits or emission targets, or:

ii Economic instruments; e.g. emission taxes, emission charges, or tradeable emission permits.

While the former types of instruments are widely implemented in environmental policy in various countries, the latter instruments are only rarely employed. In the long-lasting debate about the correct policy measures during the last two decades [Hahn 1989], the advocates of the former have been politicians, administrators, jurists and engineers. Economists, however, have been in favour of the latter because they are more economically 'efficient' (in the sense of Pareto) than the former, and therefore less economically costly in their overall operation. With rising environmental costs, the economists' arguments have gained importance. More and more, environmentalists and politicians have gradually come to favour the use of economic instruments.

To decide which instrument or policy mix should be undertaken to reduce CO_2 emissions, we need some criteria to assess the different policy options. We note the following principles, which have proved useful for practical implementation in environmental policies [Faber, Stephan and Michaelis 1989:183-187].

i Environmental policy should be economically efficient; i.e. it should attain emission targets with minimum cost, and the policy should encourage CO_2 saving techniques.

ii It should be easily possible to adopt the policy measure in cases when new knowledge about (e.g.) the greenhouse effect becomes available. From this, it follows that the policy instrument needs to be flexible.

iii Large income and wealth redistributions should be avoided as far as possible, because their occurrence always creates great resistance to the implementation of environmental (or other) policies.

iv Because of its complexity, an economic system needs much time to adapt itself to a new framework if no large frictions are to occur. Therefore new legislation should not come into force within a short time span. Instead, it should be announced well in advance and implemented only gradually. A non-environmental example of such pre-announcement of regime change is that of the implementation of the 'Single Market' in the EC, as from January 1993. This was announced many years before it was finally implemented.

v It is important to employ instruments with which a country has some experience. This is, of course, particularly the case if the country has already implemented a particular instrument. In this case the economic agents (enterprises, administrators, politicians, consumers) already have some familiarity with this means of environmental regulation, so have already developed the institutions and monitoring systems appropriate to the environmental legislation.

These five criteria need to be taken into account when one assesses different environmental policy options and instruments. For the practical implementation of CO_2 emission reduction, we believe it would be useful to use a combination of regulative laws and economic instruments. Hence, for an effective control of CO_2 emissions one needs to take recourse to economic instruments, such as tradeable emission permits or carbon taxes.

The implementation of a carbon tax has been discussed intensively in the EC, and in various individual European countries, over the past few years. Also, in Germany some experience with various charges and taxes in environmental legislation has been developed [Faber and Stephan 1987; Faber, Stephan and Michaelis 1989]. We therefore give a brief summary of the advantages and disadvantages of this particular economic instrument. (For a more detailed discussion, see Pearce [1991b].)

1. A carbon tax corrects an economic distortion which arises because of externalities, due to excessive use of environmental services. To avoid large income and wealth redistribution, it is important that the carbon tax is fiscally neutral. Since an effective carbon tax would lead to large revenues, this should be used to lower other taxes, especially those which are particularly allocatively distorting.

2. However, there also exists an advantage to the occurrence of large carbon tax revenues. It is expected that if a 'carbon convention' is to be established, the core of such a convention will need to be the allocation of reduction targets for CO_2 emissions between the different countries concerned. To persuade the less developed countries to participate in such a convention, some sort of side payments will be necessary. This could be done in a similar way as under the Montreal protocol, for the protection of the ozone layer by the

elimination of CFCs. In the latter case the developed countries have established a fund, from which the less developed countries can receive subsidies to adjust their technologies appropriately.

3. The most commonly discussed disadvantage of the carbon tax is the problem of its ecological efficiency. To reach the emission target, one has to calculate the correct corresponding tax rate. Since this cannot be calculated accurately, because of problems of information availability and modelling, the implementation of a carbon tax would not guarantee the achievement of the target emission reduction. However, in our opinion this is not a strong argument. As experience with the practical implementation of environmental taxes has shown, it is not so important to know the exactly correct tax rate in advance. There is ample evidence that it is sufficient to give the economic agents a price signal in the right direction. The dynamics of the economy will then allow the simultaneous adjustment of the instrument and the economy's response, so that the target is reached in the course of time.[8]

Finally, we note that the debates about the right policy mix are often controversial and sometimes even heated. As observers, we sometimes get the impression that the participants are so involved in the complexity of the intricate details of their discussion, that they lose sight of the aim of the debate, which is the reduction of CO_2 emissions. We regard this as regrettable. Instead we hope that economists, policy makers, and all those concerned with the well-being of future generations and the health of the global ecosystem, will recognise that debate about the precise nature of policy is only the minor component in achieving these aims. The greater component is the achievement of consensus and social 'will' that the aims *are* achieved. Let us never lose sight of this.

1.6 The Contents of the Book

This book is structured in five sections, as follows.

Part I is the *Introduction*. In Chapter 2 we outline the concepts and methods which underlie this study. In particular we stress the need to take an 'ecological economics' approach to the issue of global climate change.

---

[8] This argument is based on our experience with German environmental policy, especially water legislation.

Part II concerns *Understanding CO_2 Emissions*, and contains three chapters. Chapter 3 summarises the problem of climate change, and examines some of the approaches to its implications in the economics literature. In Chapter 4 we focus on the emission of CO_2, and seek to understand the way this has changed over time in terms of certain economic determinants. Here we introduce a method of decomposing the rate of change of CO_2 emission into various economically determined components, using the theory of central differences. The mathematics of this technique is presented in Appendix A4. The economic analysis of the determinants of CO_2 emission is continued in Chapter 5, with particular reference to Germany and the UK. The method of decomposing the rates of change of CO_2 emission by differencing is extended, to allow for changes in sectoral composition to be included.

There are three chapters in Part III, on the *Modelling Approach*. Chapter 6 outlines the modelling framework we use, which is input-output analysis, derived from the 'Activity Analysis' approach to production. In Chapter 7 we introduce input-output methods. This chapter serves either as an introduction to those unfamiliar with this technique, or a reference source for those already conversant with it. Input-output methods are extended in Chapter 8, to allow them to be applied to the use of fossil fuels, and the emission of CO_2, as attributed to various economic sectors, and various types of economic activity.

Part IV is on *Data Analysis*, and consists of two chapters. Chapter 9 presents the input-output and energy data for Germany and the UK which we use. In Chapter 10 we explore this data further, using input-output methods, and the decomposition of the changes in CO_2 emissions using difference equations. Appendix A10 contains detailed tables concerning this decomposition.

Part V concerns the analysis of *Scenarios* for reducing CO_2 emissions by Germany and the UK, and contains three chapters. Chapter 11 examines various means by which CO_2 emissions could be reduced, in terms of changing consumer demand, changing fuel mix and fuel efficiency, and changing technologies of production. In Chapter 12 we introduce a new method of establishing scenarios for reducing CO_2 emissions, which takes a mathematical programming approach. This method seeks to minimise the 'disruption' to an economy in achieving a certain target rate of reduction of CO_2 emissions, and also target rates of change in GDP and employment. This 'minimum disruption' approach to scenario analysis is applied to the German and UK data in Chapter 13.

Part VI contains one chapter, on the *Policy* implications of our study. Chapter 14 draws upon the scenario analysis in Part V to establish the likely feasibility of reducing CO_2 emissions according to the 'Toronto Target', of 20 per cent reduction over 20 years.

2 Concepts and Methods for the Study

2.1 Introduction

In this chapter we seek to lay the conceptual foundations for our study. The structure of the rest of this chapter is as follows.

Section 2.2 outlines our point of departure in this study. In Section 2.3 we characterise the CO_2 problem. Section 2.4 outlines the general principles of ecological economics which we utilise throughout the study. Our aims for this study are discussed in Section 2.5, and our method of study is outlined in Section 2.6.

2.2 Our Point of Departure

There is ample evidence that the use of fossil fuels has caused a long-run increase in atmospheric CO_2-concentrations [Intergovernmental Panel on Climate Change (IPCC) 1990]. There is also growing evidence that the emissions of CO_2, and of other trace gases, are likely to have an impact on the climate system, and this impact will be generally harmful to human activity and well-being [Cline 1991].

In the evaluation of this effect, the question often arises as to whether this effect might have catastrophic consequences:

a. For certain regions of the world.

b. For humankind as a whole.

c. For the entire biosphere.

We believe that it is not possible ex ante to establish a definite answer to these questions. We therefore propose that we put them aside. This lack of adequate answers does not imply that we should take no action.

Instead we wish to take the lawyer's notion of what seems reasonable as a conclusion to a 'reasonable human being'. This does not require evidence which is conclusive. Instead it relies upon a prudent evaluation in the light of currently available information, e.g. as presented in Chapters 2 and 3 below.

This can be extended to an attitude towards nature which does not see nature as a *given*, to be exploited. Rather we would prefer to assert a non-evaluative worth to nature. This may be approached in two ways. First, by establishing a set of 'natural values', of which market values may be a special case, or even a distortion[1]. Second, we may simply say that the natural world is valuable in itself, and just as we give consideration to the well-being of other people in society, simply because they are people, so should we give consideration to the well-being of other elements of our world, simply because they are such elements.

This notion of caring without the intermediation of any judgemental evaluation is akin to, say, the unreserved love of a mother for her child, or the unreserved love of believers for their church or god. This attitude to nature has many aspects of a religion. An alternative way of viewing it is in terms of a telos, i.e. as a goal in its own right. So we do not consider how we may achieve the 'best' return from the natural environment. Rather, we take as our telos the maintenance of the natural world as something which has an existence other than for humankind's benefit[2].

2.3 The CO_2 Problem

The nature of most environmental problems can be characterised by three features:

i Their harmful effects are generally unexpected and hence come as a surprise.

ii Response to them is therefore reactive.

iii There exist end-of-the-pipe technologies to reduce the harmful effects.

In contrast, the CO_2-problem is different, because:

a. Its bad effects, like unwanted warming of certain areas, and flooding, are anticipated.

---

[1] These 'natural values' could be determined by using a biologically founded optimisation process. The objective function is linked to the reproduction of the corresponding species, and the boundary conditions to those of the ecological system.

[2] We note, that the attitude outlined above is in line with our concept of openness [cf. Faber, Manstetten and Proops 1992a].

b. The responses planned are anticipatory.

c. There exists (so far) no end-of-the-pipe technology to reduce existing stocks of CO_2 emissions on a large scale.

The CO_2 problem, in terms of its economic analysis, is relatively straightforward, for the following reasons.

1. In industrialised countries the great bulk of CO_2 emissions result from burning fossil fuels (coal, oil and gas).

2. Industrialised countries keep very detailed records of these fuels used throughout the economy.

3. In contrast to the impacts of many other pollution problems, there is little uncertainty about the direct long-run impact of the CO_2 emissions on its atmospheric concentration (though there is considerable uncertainty about its long-run climatic effects). Here the problem is much simpler than that of, say, the leaching of pollutants from waste disposal sites into groundwater, where there is great uncertainty about the nature of the pollutants, their quantities, and their long-run effects on groundwater quality.

4. Finally, we are in no position to switch off CO_2 emissions on a global scale in a short time, as is presently being attempted for chlorofluorocarbon (CFC) production. This is because coal, oil and gas are the main sources of energy supporting modern economic activity, and it is physically impossible to carry through any production activity without energy. Since this is so, and since also the different fossil fuels are characterised by different CO_2 outputs per unit of energy delivered, we see that the CO_2 problem is not only an environmental problem, but also gives rise to an essential resource problem.

2.4 General Principles for Ecological Economic Modelling

At the end of the sixties economists reacted to the increasing recognition of environmental and of resource problems by creating the field of environmental and resource economics. Thus the *Journal of Environmental Economics and Management* was founded by the Association of Environmental and Resource Economics in 1970. To this end, concepts and instruments of neoclassical economics were used. To study environmental questions the theories of external effects, of public finance (in particular of public goods and of taxation), of property rights, of game theory, national accounting and of cost-benefit analysis were employed. To examine resource problems, mainly neoclassical capital theory and intertemporal allocation theory were used.

In spite of various valuable contributions by environmental and resource economics, several economists and representatives of other social sciences, as well as of natural sciences, felt that there was the need for a more encompassing approach. This would not rely solely on the application of neoclassical economics, but had to be of an interdisciplinary nature. To this end, the International Society for Ecological Economics was founded, with the production of the journal *Ecological Economics* since 1990. A broad definition of ecological economics is: "Ecological economics studies how ecosystems and economic activity interrelate." [Proops 1990:60]. (See also Costanza [1990] for a description of ecological economics. Also see Costanza [1991], where are published the invited papers to the First Conference of the International Society of Ecological Economics.)

Since this field is rather new, it is useful to formulate general principles which we consider to be suitable for analysing problems of ecological economics. The eight principles we espouse are discussed in the following sections.

2.4.1 Empirical Foundations

For modelling concrete problems (e.g. waste, water and, of course, CO_2 problems), one has to gain a firm knowledge of the empirical aspects of one's problem. For example, if studying water, one has to know how a water treatment plant works, what kinds of water treatment plants exist, and how they are institutionally organised[3]. This knowledge is essential for modelling, as it supplies a compass in the jungle of economic theory. For example, in the case of CO_2 we need:

1. Information about chemistry; e.g. we need to know about chemical combustion processes which generate CO_2 and other greenhouse gases.

---

[3] See e.g. Faber, Niemes, Stephan [1983] for empirical studies on water and Faber, Stephan and Michaelis [1989] and Michaelis [1991a] on waste.

2. Climatological information; we have to know about how climate is affected by CO_2 emission, and the corresponding impacts on the economy.

This shows that for ecological economic modelling a truly interdisciplinary approach is needed. In particular, the passing of time and the natural boundary conditions must be allowed for explicitly in ecological economic modelling, to which we turn next.

2.4.2 Time

The time structure of the problem should be recognised. One way to do this is to differentiate between variables which change rapidly and those which change slowly. Appropriate candidates for fast variables are flows, and for slow variables, stocks. Important economic concepts in the study of time are superiority and roundaboutness, as well as time preference [Bernholz 1971; Faber and Proops 1990:Chs. 8, 9; 1991a; 1991c]). Also irreversibilities have to be considered [Faber and Proops 1990:Chs. 4-6, Schmutzler 1991].

With respect to CO_2 emissions, we have to take account of the fact that the current state of nature is due to an accumulated flow of CO_2 in the past ('accumulated stocks matter', analogously with capital theory). These negative externalities that have been accumulated over time cannot be removed by a short-run policy, as there is no possibility of accelerating the removal of the accumulated CO_2 by economic or technological means. Of course, the short- and long-term efficiencies of policy instruments may differ.

2.4.3 Natural Boundary Conditions

By 'natural boundary conditions' we mean the nature of the world as we find it. For example, the First and Second Laws of Thermodynamics are natural boundary conditions, as are gravity, the reproductive potential of species, the availability of natural resources, such as oil, coal and iron ore, and the natural resilience of ecosystems.

For most long-run economic activities, such natural boundary conditions contribute decisively to the set of possible outcomes. Hence it is necessary to have knowledge of the relevant natural processes, and to formulate the corresponding natural constraints, as well as their interrelationships. [Faber and Proops 1985, 1990; Faber, Niemes and Stephan 1987].

Designing efficient environmental policies and environmental institutions, which explicitly take into account these boundary conditions, is one of the most complex decision making processes one can imagine. In this context, the assumption of a completely rational utility and profit maximiser is often used in the literature of environmental economics. In contrast, we consider this assumption concerning human behaviour in general too demanding as a standard. So one of the 'natural' boundary conditions which ecological environmental modelling should take account of is, instead of utility and profit maximizing, the assumption of the 'bounded rationality' of economic agents and of policy makers. However, the theory of bounded rationality as, for example, presented by Simon [1972], is , as so far developed, more an informal guide than an empirically testable theory which could be included in our formal models, to be developed in later chapters of Parts IV and V.

2.4.4 Non-Linear Relationships

Ex ante, there is a large variety of (more or less) flexible functional forms which could be used to represent a particular economic relationship. However, in general there does not exist an algebraic functional form which satisfies all plausible selection criteria. Therefore, one often assumes that locally a given non-linear function derived from economic theory can be approximated by a linear map. In our model, the assumption of linearity has the advantage of computational tractability. However, it is a crucial assumption, which requires that we consider relatively short time horizons when we use fixed coefficients, and that we evaluate our results with much care. When we examine longer-run trends we also need to assess how the coefficients themselves change over time. We shall return to this question in Chapter 6 below.

2.4.5 Production

Production processes and the means of production can be easily observed in reality. The stock of the means of production can be changed only slowly. Thus the structure of production of an economy, e.g. of the Federal Republic of Germany, is a variable which changes only slowly over time, hence it is a rather stable element of an economy. It is therefore not only easily observed, but also relatively easy modelled with input-output theory. Evidence for this are the many input-output data and studies already available.

From what has been said about time, and natural boundaries, it follows that the concept of the standard neoclassical production function [Burmeister and Dobell 1970:9-10] is only in rare cases suitable for the analysis of long-run environmental problems [Faber and Proops 1990:Ch. 8]. Instead, the use of production processes is appropriate. If these are linear, then the corresponding production functions are linear-limitational. Technical progress can also be modelled with production processes. The concept of a technique [Faber and Proops 1990:116] can be developed in a natural way (for a more detailed discussion of production and technical change see Chapter 6 below). In contrast to the standard neoclassical production function, our experience has been that the concept of a production process can be easily explained to non-economists, in particular to natural scientists and engineers.

The interplay between the natural boundary conditions on production, and the technology available, determine the set of feasible production techniques [Faber and Proops 1990:99-104]. This interaction will also engender technical progress, which will expand the technology, and cause the evolution of the set of feasible techniques. Thus the natural boundary conditions are decisive in shaping the set of possible outcomes in the long-run.

2.4.6 Demand

Since in the long-run the set of feasible outcomes is given by the natural constraints of the environment (including resource availability), it follows that the demand of an economy has always to adjust to these boundary conditions. Hence we conclude that the modelling of the interplay of boundary conditions and the technology should be given higher priority than the modelling of demand conditions for understanding long-run economy-environment interactions.

2.4.7 Quantities and Prices

From the point of view of ecological economics, it is obvious that market prices, and also the supply of public goods, are distorted. This is so because they do not encompass the true intertemporal opportunity effects of the depletion of resources and pollution which occur because of our production and consumption [Faber and Proops 1991a, 1991b]. Hence ecological economic problems should always first be analysed in terms of quantities, i.e. in terms of the primal problem. Only thereafter should one analyse the value problem of the related prices, the dual problem. Then one can look for suitable institutional implementation of environmental policies, be it via regulation or via market influencing instruments, such as taxes, charges or licenses.

2.4.8 Modelling Ex Post and Ex Ante

In general, ecological economic modelling, like any other form of economic modelling, should clearly distinguish the problem of ex ante choice from that of ex post choice. Any functional form chosen ex ante should be tested ex post before it is used to analyse economic problems. In modelling long-run economy-environment interactions, it must be acknowledged that over time, novel relationships will emerge. Once these relationships have emerged, they may be modelled ex-post. However, prior to their emergence, as they are unknown, they cannot be modelled. That is, modelling ex-ante is not possible when there is the emergence of novelty [Faber and Proops 1990:Chs. 2, 3].

Our eight principles of ecological modelling are very encompassing and demanding. We therefore cannot expect to realise them all to the same degree in this empirically oriented study. However, we should attempt this as far as possible.

Theoretical and empirical applications of various aspects of different combinations of some of these general principles are given in Proops [1977, 1984, 1988], Faber, Niemes and Stephan [1987], Faber and Proops [1985, 1990, 1991a, 1991b], Wodopia [1986a], Faber and Wagenhals [1988], Gay and Proops [1992], Faber, Stephan and Michaelis [1989], Stephan [1989], Faber, Proops, Ruth and Michaelis [1990], Michaelis [1991b], Symons, Proops and Gay [1991].

2.5 Our Aims

We wish to contribute to the analysis of the Global Greenhouse Effect both theoretically and empirically.

Though our focus in this study is on the comparison between the Federal Republic of Germany (in its borders of October 1989) and the United Kingdom, we are ultimately interested in a global approach. In contrast to many other national models and empirical studies, we shall therefore emphasise the transparency and the comprehensiveness of our approach. This we do by specifying explicitly all of our assumptions in a uniform way. We shall also determine what part of the CO_2 emissions, which occur because of goods that are exported, are attributable to the country in which they are produced and what part is attributable to the country in which they are used (see Chapters 7 and 8 below). Further, our approach will be such that it can be viewed as the nucleus of a global study.

The reasons we start with a two country comparison, instead of beginning immediately with a global approach, are the following. We wish to obtain a thorough grounding in the economies of two important countries of the global economy. This will give us a firm understanding of their structures, at both disaggregated and aggregated levels. By studying the two economies over many periods, we shall get to know how they evolve over time. The knowledge gained in this way is of the utmost importance for policy conclusions and implementations, at both national and international levels.

The essential prerequisite for this endeavour is to gather data and analyse empirically the performances of economies, within a consistent framework, and over long periods of time. This will enable us to understand economic trends and their underlying economic behaviour. This in turn will allow us to find out what, and how, aspects of reality may be modelled for the European Community, and for the world at large. This knowledge can then be employed for scenario analysis[4] and for policy recommendations.

Finally, we note that our approach will be formulated in such a way as to permit short-, medium- and long-run analysis.

2.6 Our Method

The outcome of economic activity, in particular the level and the composition of final demand, and thus in turn of gross domestic product (GDP), is driven by:

   i Short- and medium-run preferences, determined according to social conventions.

  ii The technology embodied in the capital structure of the economy.

 iii Laws and institutions.

  iv The short- and medium-run availability of natural resources.

---

[4] On the scope and the limits of prediction in economics, see Faber and Proops [1990:Ch. 3].

Increasing awareness of environmental issues has started to bring a change in economic activity. This is most clearly to be seen in the current concern over the global greenhouse effect, the depletion of the ozone layer, and the destruction of tropical rainforests. From this it follows that, while up to now short- and medium-term considerations have prevailed, there is a growing interest in long-run trends of the economic determinants of these environmental issues.

We can regard the satisfaction of culturally and historically determined needs as being the 'end' of economic activity, and we can regard the exploitation of the natural environment as one of the primary means to achieve this 'end'. The linkage between the end and the natural environment is mediated by the production structure (i.e. the available technology and the stock of capital goods), through resource use and pollution.

In the short- to medium-run, we might suppose that the culturally and historically determined needs of society are relatively unchanging. Also, in the short- to medium-run the techniques of production available, and the capital goods embodying these techniques, would also be relatively constant. On the other hand, interaction with the natural environment in the short- to medium-run has often been viewed as a source of flexibility of economic activity, as neither resource shortage nor environmental degradation through pollution were felt as constraints.

When the long-run is considered, the situation is seen to be quite the contrary to that in the short- to medium-run. Regarding the production structure, this can be altered drastically, through the invention of new techniques, their innovation through the formation of new capital goods, and the elimination of old techniques. On the other hand, the natural environment comes to be seen as a constraint on human activity in the long-run, through resource depletion and the deleterious effects on human welfare of various pollutants. Finally, and perhaps most contentiously, in our view human 'needs' are potentially extremely flexible in the long-run. Human needs derive ultimately from the human will, and through a growing recognition that the present mode of consumption is neither sustainable, nor even 'satisfying', we are confident that the ends that are willed shall, and must, change dramatically.

Despite the primacy of the will in the determination of the human impact on the natural environment, this study will not concern the will directly. Instead we focus on the means by which the will is realised; i.e. the production structure. By the production structure we mean the set of available techniques of production (the technology), and the available capital goods which use the technology, to allow production of consumption goods. We recall that in the long-run, invention of new techniques and the building up of corresponding stocks of new capital goods, allows the relationship between the end of consumption and the means of environmental exploitation to be altered.

2.6 Our Method

Now the essence of capital accumulation is the acceptance of less consumption in the present to allow *different*, and more satisfying, consumption in the future. In the past, 'different' has usually been taken to mean 'more', and where it has been truly different, this has tended to further add to the increasing environmental degradation resulting from more consumption. This doubly negative outcome for environmental degradation and resource depletion is not a necessary result of technical change, nor is it beneficial in the long-run. We believe that equating 'more future consumption' with 'different future consumption' will soon be recognised as both conceptually mistaken and unfeasible in practice. We acknowledge that this strong statement may not find favour with some economists. We assert it nonetheless.

To analyse the long-run behaviour of the technology of industrialised countries, we are going to focus on the three main sources of structural change in the foreseeable future:

1. The level and structure of the final demand for goods and services.

2. The quantity and the mix of fossil fuels used.

3. The change in the structure of the production processes that use the fuel mentioned in (2) to satisfy the demands in (1).

If we wish to change the world (i.e. to reduce the greenhouse effect) we need first to understand it. For this reason the study of the long-run behaviour of the technology is of importance. In particular we need to know how structural change occurs.

Part II

Understanding CO_2 Emissions

3 The Problem of Climate Change and CO_2 Emissions

3.1 Introduction

In Chapter 1 we outlined the problem of climate change resulting from the anthropogenic greenhouse effect. In this chapter we begin our economic analysis of the causes of the emission of the principal greenhouse gas, carbon dioxide (CO_2).

In Section 3.2 we discuss the nature of the problem. Section 3.3 reviews the evidence on greenhouse gas emissions. We review the current economic literature on CO_2 modelling in Section 3.4. Section 3.5 contains our conclusions and outlook.

3.2 The Nature of the Problem

Over the past decades we have observed an increasing atmospheric concentration of carbon dioxide (CO_2) and other trace gases, such as methane (CH_4), nitrous oxide (N_2O), chlorofluorocarbons (CFCs) and tropospheric ozone (O_3). A common feature of all these gases is that while they do not impede the passage of short-wave (visible) radiation through the atmosphere, they absorb most of the thermal infra-red radiation emitted from the surface of the earth. This absorbed radiation is re-emitted, both back to earth and out to space. Therefore we can describe the role of the trace gases as analogous to the glass in a greenhouse, and hence call these gases 'Greenhouse Gases' (GHGs), and the temperature rise they engender the 'Global Greenhouse Effect'.[1]

This greenhouse effect leads to a global mean temperature of the earth's surface of $15°C$ instead of about $-19°C$, which it would be without the existence of any of the greenhouse gases. Thus the natural greenhouse effect is essential for life on the earth. However, we have to expect a further increase in the global mean surface temperature of the earth of $2-5°C$, if anthropogenic (human produced) greenhouse gas emissions double within

---

[1] A brief summary of the scientific basis for the greenhouse effect is given in Cline [1991]. For an extended discussion, see IPCC [1990].

the next 50-100 years. The latter is possible, if the emissions of greenhouse gases by anthropogenic sources continue to grow at the present rate [Hasselmann 1991:171].

This expected global warming will affect our environment in various ways. In climate research it is widely accepted that global warming could raise the sea level between 31 and 110cm [Cline 1991:915], and that horizontal shifts in cropping patterns are possible, of several hundred kilometres for each degree Celsius of temperature change. Furthermore, a change in climate could have serious effects on global ecosystems, and therefore on growing conditions for agriculture and forests. These illustrations are only brief impressions of possible results of global warming. For a detailed discussion, see Bolin et al. [1986:Chs. 7-10].

Of the greenhouse gases mentioned above, CO_2 is the most important. It is responsible for about 55% of the anthropogenic greenhouse effect [IPCC 1990]. Its main source is fossil fuel burning, which at present is the predominant means of energy production in industrial economies. Further, energy is essential for economic development. Thus we concentrate in our study on CO_2 production and the impacts of CO_2 reduction on economic growth. But before we start with the analysis of the data concerning CO_2 emissions and gross domestic product (GDP), over time and for different regions, we briefly discuss the trends in greenhouse gas emissions.

3.3 The Evidence

In the following sections we illustrate the relationship between emissions of anthropogenic greenhouse gases and economic development.

3.3.1 Global Greenhouse Gas Emissions

In this section we illustrate the development of greenhouse gas emissions.

In Figure 3.1 we show the total emissions of CO_2 (billion tons) from the burning of fossil fuels, in the period 1860-1982. It can be seen that the emissions from burning of liquid and gaseous fuels have increased rapidly, while those from solid fuels have increased more slowly over the last decades.

Figure 3.2 summarises the production of methane from its main sources. As Bolin et al. [1986:168] discuss, total methane emissions have increased during the last four decades, at about 1% per annum. The highest rate of increase was observed between 1960 and 1975, because of a rapid increase in cattle production, rice cultivation and distribution of natural gas. In recent years, methane emissions from waste disposal have become more important, a tendency which can be expected to continue.

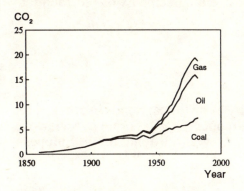

Figure 3.1. Global CO_2 emissions by fuel (B tonnes).
Source: Rotty and Masters [1985:70].

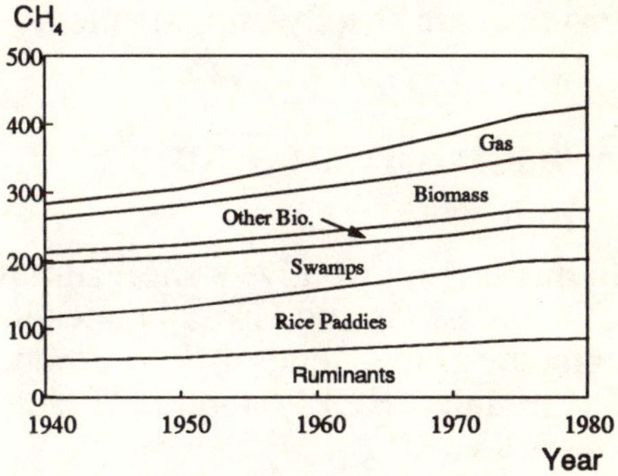

Figure 3.2. Global methane emissions by source (M tonnes).
Source: Bolin et al. [1986:167].

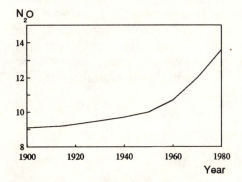
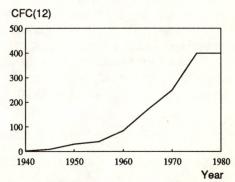

Figure 3.3. Global N_2O emissions (M tonnes).
Source: Bolin et al. [1986:174].

Figure 3.4. Global CFC(12) emissions (K tonnes).
Source: Bolin et al. [1986:175].

Figure 3.3 shows that the rate of growth of nitrous oxide (N_2O) emissions has increased from about 0.1% per annum at the beginning of this century, to about 1.3% p.a. during the last decade. This is due to increasing N_2O emissions from fertilizer application and the combustion of fuels [Bolin et al. 1986:173].

Chlorofluorocarbons (CFCs) are solely anthropogenic in origin. Figure 3.4 shows the historical emissions of the most important of these [CFC(12)] since the beginning of its production. It is expected that CFC emissions will decrease in the coming decades, if the agreements of the Montreal Protocol are strictly implemented.

3.4 Historical Trends

In this section we give a short summary of historical trends in energy use, CO_2 emissions and Gross Domestic Product (GDP). For an overview we compare global trends in these components with the trends for the European Community (EC12)[2] and the USA.

3.4.1 Conversion Units

The first step in assessing fuel use, and the resulting CO_2 emissions, is to establish the quantities of these fuels used, and the quantity of CO_2 emitted for each unit of fuel burnt. This information is summarised in Table 3.1. There we indicate the units in which the original data were collected. The conversion to the common unit of 'tonnes of coal equivalent' (tce) was achieved using the ratios also shown in that table. We then indicate the proportion, by weight, of carbon in each fuel. Finally, the quantity of CO_2 emitted per unit of fossil fuel burnt is calculated.

Fuel	Units	Units/tce	% Carbon	CO_2/tce
Coal	tonnes	1.00	0.60	2.20
Oil	tonnes	0.59	0.84	1.82
Gas	kTherms	0.25	0.75	1.35
Nuclear/hydro	TWh	2.43	-	-

Key: kTherms - 10^3 therms; TWh - 10^9 watt hours; tce - tonnes of coal equivalent.

Table 3.1. Fuel units and conversion ratios to CO_2 emissions.

---

[2] Because the number of members of the EC changed during this period, we treat the EC as consisting of the present 12 member states for the whole time period.

We can use the data in Table 3.1 to compare the CO_2 per unit energy for oil with that for coal. For oil, we generate 1.82 tonnes of CO_2 per tce, while for coal the figure is 2.20 tonnes of CO_2 per tce. So the CO_2 per unit of energy for oil is 1.82/2.20 = 0.827 = 82.7%. This high figure is mainly because of the relatively low proportion of carbon in coal. A surprisingly large proportion of coal (an average of 40% by weight) is silica, water, and various other impurities.

3.4.2 Data on Energy Use, CO_2 Emissions and GDP

Figure 3.5 and 3.6 show energy use by the World, the USA and the EC, both in absolute terms, and as indexes from the 1950 figures. We see that energy use by the World and the EC grew significantly faster than that by the USA in 1950-73. Following the impact of the first oil price 'shock', energy use by the USA became approximately steady, while the EC energy use showed an overall decline up to 1988. However, World energy use continued to grow between 1973 and 1988.

In Figures 3.7 and 3.8 are the CO_2 emissions by the World, the USA and the EC, again in absolute and index form. The temporal pattern of the emission of CO_2 is similar to that for energy use, discussed above.

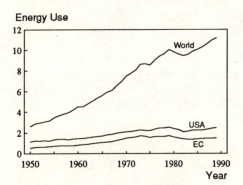
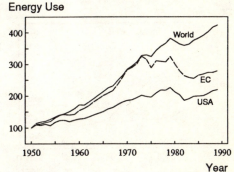

Figure 3.5. Energy use (B tonnes coal equivalent): World, USA and EC.
Source: UN [1976], OECD [1989].

Figure 3.6. Energy use (index): World, USA and EC.
Source: Figure 3.5.

From Figure 3.7 we see that global CO_2 emissions have been on a steadily upward trend, apart from a brief 'dip' in the early eighties. In contrast, from Figure 3.8 we see that CO_2 emissions by the developed world, as represented by the EC and the USA, has had a much less rapid rate of growth. Indeed, for the EC the peak in CO_2 emissions was reached in the mid-seventies, with a fluctuating decline since, although the current trend

38 3 The Problem of Climate Change and CO_2 Emissions

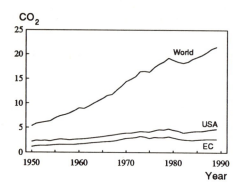

Figure 3.7. CO_2 emissions (B tonnes): World, USA and EC.
Source: UN [1976], OECD [1989].

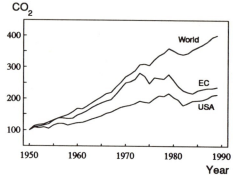

Figure 3.8. CO_2 emissions (index): World, USA and EC.
Source: Figure 3.7.

is upwards. The rate of increase of CO_2 emissions by the USA has been less than for the EC, but has been more or less steadily positive, apart from a dip in the early eighties.

Although the patterns of energy use in figures 3.5 and 3.6 are similar to those for CO_2 emissions in Figures 3.7 and 3.8, in fact there have been significant alterations in the CO_2/energy ratios involved, as shown in Figure 3.9.

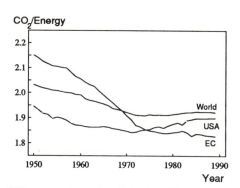

Figure 3.9. Changing energy/CO_2 ratios (tonnes CO_2/tce): World, USA and EC.
Source: Figures 3.5 and 3.7.

We see that the World CO_2/energy ratio has fallen slightly since 1950, representing a global substitution from coal to oil and gas, and the growth of nuclear power. As detailed above, the combustion of oil and gas releases less CO_2 per unit of energy generated. The trend for the USA is less marked than for the World as a whole. Indeed, since the early seventies the trend is upward, reflecting the increased use of coal for generating electricity. In

3.4.2 Data on Energy Use, CO_2 Emissions and GDP

the EC, the trend in the CO_2/energy ratio is downwards. This reflects the rapid shift from coal, to oil and gas, in these economies, and the growth in the use of nuclear power plants.

We now consider the relationship between energy use and the production of national product. Figures 3.10 and 3.11 show energy/GDP and CO_2/GDP ratios for the World, USA and EC. (GDP is measured in 1985 ECUs.)

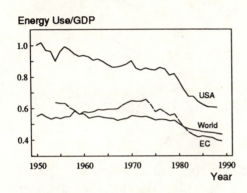
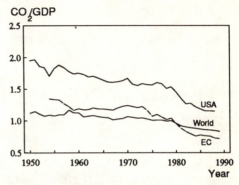

Figure 3.10. Energy/GDP (tce/K ECU): World, USA and EC.
Source: UN [1976], OECD [1989].

Figure 3.11. CO_2/GDP (tonnes/K ECU): World, USA and EC.
Source: Figure 3.9.

Both the energy/GDP and CO_2/GDP ratios are generally falling, for the World, the USA and the EC. The declines in the energy/GDP ratios are steeper than for the CO_2/GDP ratios, reflecting the changes in CO_2/energy ratios discussed above. This general decline in the CO_2/GDP ratio, which is particularly apparent for the USA and the EC, is a hopeful sign that the reduction of CO_2 emissions by the developed countries is achievable. This reduction may be attributed to three causes: changed fuel mix; improved 'efficiency' of energy use; and changed mix of output by the economy. These three aspects of the CO_2-GDP relationship will be explored in more detail in Chapters 4 and 5.

Before we move to a review of CO_2 modelling in the economic literature, we consider the distribution of CO_2 emissions between different economic regions. Figure 3.12 shows that during the last four decades the share of CO_2 emissions produced by the USA and Western Europe decreased. Conversely, we observe an increasing contribution from the developing countries to global CO_2 emissions. This is due to the growing industrialization in these countries, and we can expect this trend to continue.

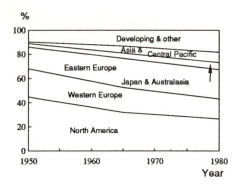

Figure 3.12. Changing patterns of CO_2 emissions.
Source: Rotty and Masters [1985:77].

3.5 Review of CO_2 Modelling in the Economics Literature

As described in the previous section, we are confronted with a widespread range of consequences, if global warming takes place. For instance, average sea-level rise, a change in average regional temperature and rainfall, and changes in the frequency and severity of weather events and natural disasters [Barbier and Pearce 1990:11-12]. In the face of these possible impacts of global warming, there are two feasible policy responses.

1. The so-called 'do nothing' approach. We do nothing now and adapt to the impacts as they arise.

2. A preventive approach. We attempt to avoid possible future damage, by reducing greenhouse gas emissions now and in the future.

Between these two different approaches every combination of prevention and adaptation is possible. These combinations lead to different costs, both in absolute amount and in temporal distribution. Thus an economic decision as to which policy should be implemented, requires the calculation of costs and benefits of different policy options. However, this is not a simple task, as there is great uncertainty about future events. So it is difficult, even impossible, to calculate future costs, especially if we are confronted with irreversible and unknown events and developments.

Given these difficulties, economic modellers typically proceed as follows. They first plot paths for greenhouse gas emissions and gross domestic product (GDP) in the absence of abatement policy. This constitutes the 'base line scenario'. They then plot alternative paths, embodying policy measures, in the form of abatement or adaptation (e.g.

building dykes and employing different tilling techniques in agriculture). These constitute the policy scenarios. The costs of controlling GHG emissions are measured as the deviation of GDP under abatement from its base values. However, one has to be aware that these costs, measured in terms of GDP losses, do not adequately reflect changes in social welfare, since the GDP is not an appropriate welfare measure [Daly and Cobb 1989; Norgaard 1989; Faber and Proops 1991a, 1991b].

The method of modelling described above, is called a 'top-down' approach. Darmstadter [1991:8] characterises these models as macroeconomic-technological models, linking energy and economy. In these models three types of inputs are typically distinguished: labour, capital, and energy. The time horizons of these kind of studies are long, between 40 and 110 years. From this it follows that their authors are confronted with inherent uncertainties concerning the key factors of future developments of the economy, such as population growth, changes in energy efficiency, the availability of techniques to exploit solar energy, etc., to be discussed below.

In contrast to this highly aggregated economic modelling is the 'bottom-up' approach. This focuses on those consumption and production activities in the economy which use energy. It analyses what possibilities exist to conserve energy, e.g. by implementing energy saving technologies. Top-down modelling and bottom-up modelling are not exclusive of each other, but describe two important perspectives in analysing the costs of reduction of GHG emissions. Before discussing the possibilities of integrating these two approaches we look at those key factors which determine future CO_2 emissions.

3.5.1 Factors Determining CO_2 Emissions

In most CO_2 reduction scenarios, an increase or reduction in the level of emissions is determined by assumptions about:

1. Output and population growth.

2. Energy efficiency.

3. Energy price.

4. The availability of so-called back-stop technologies (e.g. solar power plants).

We now examine these factors in turn.

3.5.1.1 Output and Population Growth

As mentioned in the review article by Hoeller et al. [1991:52], most studies assume a growth of GDP of around 2% per annum during the time horizon they use. Usually it is assumed that the growth in the next few decades is greater than growth in the following decades, because of lower population growth in developing countries in later decades. Furthermore, it is supposed that growth rates in developing countries are higher than in developed countries. It is important to note that all baseline scenarios reviewed by Hoeller et al. [1991] do not contain estimates of costs or benefits which arise because of climate change. Only Nordhaus [1991] has estimated damage and cost functions associated with climate change.[3]

3.5.1.2 Energy Efficiency

The rate of improvement of energy efficiency is another important exogenous parameter in the models. Estimates range between 0% and 2% per annum for different regions [Manne and Richels 1990]. The long-term average global rate for the whole world is estimated to be around 0.8-1% per annum. The rate of change of energy efficiency is an indicator of the change of an important variable, the energy/GDP ratio.

3.5.1.3 Energy Price

In most of the studies reviewed by Hoeller et al. [1991:52ff.], increases in relative energy prices are taken into account. This reflects the depletion of scarce resources and increasing extraction costs, and leads to a projected sharp rise in fossil fuel prices. Therefore, rising energy price and energy efficiency lead to slower growth of energy demand than of output growth.

---

[3] Morgenstern [1991:140] mentioned, that Nordhaus "... is the only economist intrepid enough to write about both damage and cost functions associated with climate change." For another commentary of this work by Nordhaus, see Daily et al. [1991].

3.5.1.4 Back-Stop Technologies

When the price of the non-renewable resource becomes sufficiently high, then one may move to the use of a 'back-stop technology', which expensively employs some effectively non-depletable alternative resources; e.g. solar energy is a back-stop technology for fossil fuels.

Most studies contain an exogenous reduction in prices in the course of time for back-stop technologies, due to technical progress. This in turn brings about a decrease in the CO_2/energy ratio.

3.5.2 Key Determinants of the Costs of Abating GHG Emissions

Relating to the factors discussed above, we can identify the following four categories of costs:

1. Costs because of substitution possibilities for a given technology of an economy between:

   a. Energy and other production factors.

   b. Different fuels in energy production.

   c. Production factors differing in energy intensity.

   d. Final goods differing in energy intensity.

2. Costs due to technical progress and innovation in:

   a. Energy-use technology.

   b. Production technology (inter-industry trading).

   c. Non-fossil fuel generation (or back-stop technology).

2. Costs due to changes in relative prices:

   a. Changes in the terms of trade.

   b. Changes in energy prices without CO_2 emission abatement.

4. 'Synergetic'[4] effects:

   a. Reduction of CO_2 emissions implies a reduction in other emissions, e.g. SO_2 or N_2O.

   b. Elimination of inefficient technologies.

3.5.3 Emission Reduction Scenarios

The great differences in the estimates of most studies result from the use of different assumptions about the factors described in Section 3.5.1 above, reflecting:

1. The fact that there is no agreement on the constrained level of GHG emissions.

2. The non-uniformity of the time profile of reduction of CO_2 to be modelled.

3. The use of different time horizons.

3.5.3.1 Reduction Targets

Manne and Richels [1990] look at a time horizon from 1990-2100. The emission target is to reduce CO_2 emissions by 20% from the reference year by 2020, and keep this level until 2100. In contrast to this scenario, Dixon et al. [1989] simulate the implications of the Toronto target. This implies a reduction of CO_2 emissions by 20% below the level of 1988 by 2005. The implications of this reduction goal are also discussed in a dynamic general equilibrium model for the U.S. congressional budget office [CBO 1990].

---

[4] A synergetic effect occurs when the outcome of the influence of simultaneous factors is different from the sum of the two effects that are caused by each factor on its own [Victor 1972:230].

3.5.3.2 Policy Instruments

The main policy instruments considered in the CO_2 models are carbon taxes. These taxes are differentiated by the carbon content of the different types of fossil fuels. A discussion of the advantages and disadvantages of a carbon tax in adjusting to global warming, is given in Pearce [1991b].

Further instruments for the reduction are tradeable emission rights. As Manne and Richels [1991:138] mentioned in an article about international tradeable carbon emission rights, this leads to lower costs of CO_2 reduction, in comparison with a carbon tax.

3.5.3.3 Growth Effects

The review of Hoeller et al. [1991:57] shows that in most studies the reduction of CO_2 emissions leads, in the long-run, to a reduction in the growth rate of GDP, of up to 0.2% per annum. This implies that the GDP growth rate is about 1.8% annually in the abatement scenario, as oppose to 2% annually in the baseline scenarios[5].

In Part V below, we shall develop emission reduction scenarios for our own study.

3.6 Outlook and Conclusions

In the introduction to this section, we distinguished between top-down and bottom-up modelling. One possibility to integrate these two perspectives of modelling CO_2 reduction is our input-output approach. Starting with 47 sector input-output tables, for different years, we are able to model the production structure of an economy. Further, the input-output tables contain all necessary macroeconomic aggregates (e.g. GDP), so that we are able to create scenarios, as in the top-down models discussed above.

Furthermore, the difference equations we develop in Chapter 4 allow one to relate changes in CO_2 emission to factors such as improving energy efficiency, changes in fuel mix, and changes in the structure and/or size of final demand. These give detailed insights, as in a bottom-up analysis.

---

[5] A detailed comparison and discussion of growth effects, tax rates and CO_2 reduction is given by Boero, Clarke and Winters [1991:Ch. 4].

Finally, we are able to widen the approach by combining the input-output model with an expenditure system, such as the SPIT model for the U.K. [Symons, Proops and Gay 1991].

4 Decomposing the Rate of Change of CO_2 Emissions

4.1 Introduction

In this chapter we seek to understand more fully the data presented in Chapter 3. We do this by first deriving a technique of decomposing the proportional rate of change of CO_2 emission into three main factors. We then apply this technique to our data for the World, the USA and the EC.

In more detail, the structure of this chapter is as follows. Section 4.2 considers how the time structure of CO_2 emissions might be understood. In Section 4.3 we consider how CO_2 emission can be represented as the product of several economic variables. Section 4.4 shows how this multiplicative relationship can be decomposed by differencing. This differencing approach is applied in Section 4.5, to examine the rates of change of CO_2 emissions by the World, the USA and the EC. Conclusions are drawn in Section 4.6. (A more detailed and general approach to 'decomposition by differencing' is given in Appendix A4.)

4.2 Understanding the Time Structure of CO_2 Emissions

Having examined the historical behaviour of CO_2 emission by the World, the USA and the EC, we now wish to understand *why* this behaviour occurred. Such an understanding requires, of course, an *economic model* of CO_2 emissions.

In Figure 3.7 we saw that World CO_2 emissions have increased since 1860. This is also true for World Gross Domestic Product (GDP). Thus up to now there has been a strong correlation for the World between CO_2 emissions and GDP.

Now in Figure 3.11 we saw that the relationship between CO_2 emission and national income seems to be reasonably well described by a ratio which steadily declines over time. However, we should not be satisfied that this is sufficient to allow modelling and policy prescriptions. We know that different fuels generate different amounts of CO_2 for the provision of the same amount of useful energy. We also know that different industries have different energy requirements and use different fuel mixes. Further, we know that the mix of industries itself changes over time. Finally, the

48 4 Decomposing the Rate of Change of CO_2 Emissions

efficiency of fuel use varies between industries, and over time. Thus the emission of CO_2 by an economy will necessarily reflect a range of influences, that will vary independently over time.

4.2.1 The Influential Variables

We begin our assessment of the factors influencing CO_2 emissions by considering aggregate economic activity. That is, in the first instance we shall not be concerned with the structure of industrial output, in terms of the relative sizes of the industrial sectors, and the techniques of production they employ. (We shall turn to this problem in Parts III-V; indeed, the bulk of this study is devoted to precisely the issue of the importance of the various industrial sectors in understanding and modelling CO_2 emissions.)

The first step in seeking to understand the time path of CO_2 emissions by economies is to establish the variables that we feel to be influential. For reasons that will become apparent shortly, we identify the following as being influential variables:

1. The ratio of CO_2 emission (C) to the corresponding energy use by the economy (E); i.e. we consider (C/E).

2. The ratio of energy use by the economy (E) to the Gross Domestic Product of the economy (Y); i.e we consider (E/Y).

3. The Gross Domestic Product of the economy (Y).

4.3 CO_2 Emissions as a Product of Several Variables

Rather than relating these variables in a functional form, which we could estimate with the normal econometric techniques, we take a different, and we feel novel, approach. We begin simply by noting the identity:

$$C \equiv \left(\frac{C}{E}\right)\left(\frac{E}{Y}\right) Y$$

We therefore identify the total CO_2 emission in a year as being influenced in three ways.

First, the mix of fossil (and non-fossil) fuels may alter, so that the same amount of energy is provided, but a different amount of CO_2 is emitted (i.e C/E may alter). This assumes that (E/Y) and Y are unchanged; i.e. the usual economists' assumption of *ceteris paribus*. For example, the switch in the UK economy away from coal and towards gas would have reduced the emission of CO_2 for the provision of the same amount of useful energy.

Second, the energy requirement to produce a unit of value in the economy may alter, *ceteris paribus* (i.e. E/Y may alter). For example, economies may become more efficient in their use of energy. Alternatively, the economy may evolve so that there is a shift away from the production of energy intensive goods (e.g. steel) and towards the production of less energy intensive ones (e.g. financial services).

Third, the GDP of the economy may alter, *ceteris paribus* (i.e. Y may alter).

4.4 Decomposing the Components of CO_2 Emission Changes

We recall our identity relating total CO_2 emission to the three variables discussed above:

$$C = \left(\frac{C}{E}\right)\left(\frac{E}{Y}\right)Y. \tag{4.1}$$

Now as the identity involves only multiplication, one approach is to take logarithms of the identity, giving:

$$\ln C = \ln\left(\frac{C}{E}\right) + \ln\left(\frac{E}{Y}\right) + \ln Y. \tag{4.2}$$

If we differentiate (4.2) with respect to time, we get:

$$\frac{\dot{C}}{C} = \frac{(\dot{C/E})}{(C/E)} + \frac{(\dot{E/Y})}{(E/Y)} + \frac{\dot{Y}}{Y}. \tag{4.3}$$

The 'dot' notation in (4.3) represents the first derivative with respect to time; i.e.:

$$\dot{x} \equiv \frac{dx}{dt}.$$

We see that equation (4.3) represents the proportional rate of change of CO_2 emissions $\dot{C}/C$. That is, the rate of change of C, as a proportion of its current value. This proportional rate of change of CO_2 emissions is decomposed into three components in equation 4.3. These are the proportional rates of change of the CO_2/energy ratio, the energy/GDP ratio, and GDP. For example, if C/E is falling at 1% per annum, E/Y is falling at 0.5% per annum, and Y is growing at 2.5% per annum, then CO_2 emissions are growing at:

$$-1\% - 0.5\% + 2.5\% = 1\%.$$

Now we know that none of these ratios is stationary over time (i.e zero proportional rates of change). Neither do they display completely constant proportional rates of change. It would therefore be very useful if we could plot the proportional rates of change of these elements, using our data for the World, the USA and the EC, discussed in the previous chapter. To achieve this we need to find a suitable discrete-time approximation to the continuous-time expression in equation (4.3). We can do this as follows. First, write our derivative as approximated by a difference:

$$\dot{x} \equiv \frac{dx}{dt} \approx \frac{\Delta x}{\Delta t}. \tag{4.4}$$

If we suppose that our time unit is always unity (i.e. $\Delta t = 1$), then from (4.4) we have:

$$\frac{\dot{x}}{x} \approx \frac{\Delta x}{x}. \tag{4.5}$$

Substituting the type of expression in (4.5) into (4.3) we get:

$$\frac{\Delta C}{C} \approx \frac{\Delta(C/E)}{(C/E)} + \frac{\Delta(E/Y)}{(E/Y)} + \frac{\Delta Y}{Y}. \tag{4.6}$$

This looks very promising, as it will allow us to assess trends in the components of the rate of change of CO_2 emission, and thereby make assessments of the likely behaviour of CO_2 emissions overall. In the next section we apply this decomposition method to the rate of change of CO_2 emissions for the World, the USA and the EC.

4.4.1 Necessary Extensions to the Technique

However, the approach above needs extension, for three reasons.

The first problem is that the analysis using logarithms is only effective when products are used. Later in this study we shall need to apply a decomposition of changes when we use sums as well as products in our basic identity.

Second, we have written expression (4.6) as an approximation, with no indication as to the likely size of the discrepancy between the left- and right-hand sides. If our technique is to be soundly based, we need to find some means of representing this difference, so that it can be assessed.

Third, if one differences annual data, there is often enough random variation in the data to cause the difference plot to be severely fluctuating. This is because when one differences a time series, the effect is to remove linear trends, but leave any (non-autocorrelated) error term largely unaltered. So differencing means that more of the total variation of the variable from its mean must be attributed to the error term.

All three of these problems are discussed in detail in Appendix A4, where a more general 'differential' approach to decomposition is taken, and the precise nature of the corresponding difference equations are explored.

4.5 The Rates of Change of CO_2 Emissions

Having derived a technique of decomposing the proportional rate of change CO_2 emission, we now apply it to the data established in Chapter 3, for the World, the USA and the EC. These decompositions are shown in Figures 4.1-4.3. (We use an 'eight-period difference', as discussed in Appendix A4, Section A4.6.)

These graphs each plot four curves, which are the time paths of the elements of the four components of equation (4.6). The relationship between the elements of the equation and the curves, as marked, is as follows:

$$'C' \Leftrightarrow \frac{\Delta C}{C}$$

$$'(C/E)' \Leftrightarrow \frac{\Delta(C/E)}{(C/E)}$$

$$'(E/Y)' \Leftrightarrow \frac{\Delta(E/Y)}{(E/Y)}$$

$$'Y' \Leftrightarrow \frac{\Delta Y}{Y}.$$

When the last three lines ('C/E', 'E/Y' and 'Y') are summed vertically, they give the first line ('C').

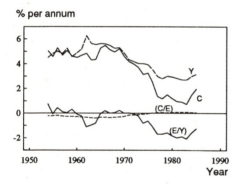

Figure 4.1. Decomposition of rate of change of World CO_2 emissions.

From Figure 4.1, we see that the proportional rate of change of World CO_2 emission is positive throughout the period. That is, CO_2 emission has been increasing. However, the rate of increase has fallen from about 5% per annum up to 1970, to 2% by the mid-1980s. This fall can be attributed to a fall in the rate of growth of World GDP, from 5 % per annum to 3.5% per annum, and a fall in the rate of growth of the energy/GDP ratio, from approximately 0.5% per annum to -1.5% per annum. The CO_2/energy ratio has been almost unchanging throughout the period, for the World as a whole.

Globally, the rate of increase of CO_2 emissions is positive, even though the 'efficiency' with which energy is converted into GDP is continuing to improve.

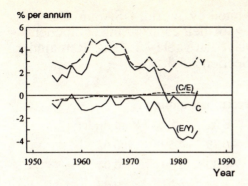

Figure 4.2. Decomposition of rate of change of USA CO_2 emissions.

Considering Figure 4.2, we see that the proportional rate of change of CO_2 emission by the USA has fallen from a peak of 4% per annum in the mid-1960s, to about -1% per annum by the mid 1980s. This fall can be attributed to a fall in the rate of growth of GDP, but particularly to a sharp fall in the rate of change of the energy/GDP ratio after the mid 1970s. The CO_2/energy ratio has remained almost unchanged.

This finding is encouraging, suggesting that the USA is capable of achieving our 'Toronto target' of a sustained 1% p.a. reduction in CO_2 emissions.

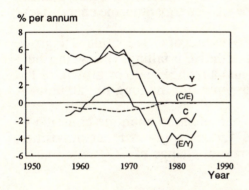

Figure 4.3. Decomposition of rate of change of EC CO_2 emissions.

From Figure 4.3 we see that the EC has also shown a fall in the rate of growth of CO_2 emission, from a peak of 6% per annum in the mid-1960s, to -2% per annum in the mid-1980s. As for the USA, this can be attributed to a fall in the growth rate of GDP, and a fall in the rate of change of energy/GDP ratio. In the EC, the CO_2/energy ratio was falling at about 1% per annum up until the late 1970s, when it ceased to alter significantly.

This is very encouraging, as the current rate of fall of EC CO_2 emissions (-2% p.a.) is *twice* that required to attain our 'Toronto' target.

4 Decomposing the Rate of Change of CO_2 Emissions

As a summary of the changes exhibited by World, USA and EC CO_2 emissions, these have been calculated, with their contributing components, for the period 1950-1990, and are displayed in Table 4.1. (The 'Remainder Term' is discussed in Appendix A4, Section A4.5.)

	\multicolumn{5}{c}{Percent per annum}				
	C	C/E	E/Y	Y	Remainder
World (1950-89)	3.83	-0.19	-0.79	4.59	0.02
USA (1950-88)	2.79	-0.01	-0.30	3.09	0.01
EC (1953-89)	1.94	-0.51	-1.64	4.02	0.07

Table 4.1. Decomposition of CO_2 emissions changes: World, USA and EC.

4.6 Conclusions

In this chapter we have further analysed the data from Chapter 3, on CO_2 emissions as related to energy use and GDP. We have derived a technique for decomposing the rate of change of CO_2 emission into components determined by the rates of change of the CO_2/energy ratio, the energy/GDP ratio, and GDP.

We have seen that while the World emission of CO_2 continues to grow, that by the USA has stabilised, and that by the EC is falling. This we attribute to a reduced rate of growth of GDP, and a reduced rate of growth of the energy/GDP ratio. This is very encouraging, for the long-run attainment of the 'Toronto target'.

In the next chapter we shall focus our attention upon the emission of CO_2 by Germany and the UK, and further apply our decomposition technique to the determinants of the rate of change of CO_2 emissions in these two countries.

A4 Appendix: Decomposition with Differencing

A4.1 Introduction

In this Appendix we derive the results on the decomposition of rates of change of a variable which depends on other variables. This is a generalisation of the method of logarithmic differentiation used in Chapter 4. We use this technique of differencing extensively in Chapter 5, and Sections III and IV.

A4.2 The Differential Approach to Decomposition

In this Appendix, instead of the logarithmic differentiation approach used in Chapter 4, we use total differentiation as our basic tool. From elementary calculus we know that we can find the total derivative of a variable, in terms of the partial derivatives of the variables upon which it depends. For example, we may have:

$$x = x(a, b). \quad (A4.1)$$

Differentiation of (A4.1) gives:

$$dx = \frac{\partial x}{\partial a} da + \frac{\partial x}{\partial b} db. \quad (A4.2)$$

Now if the function is simply multiplicative, then we have:

$$x = ab. \quad (A4.3)$$

The total derivative of (A4.3) is:

$$dx = da.b + a.db.$$

This approach works equally well if three or more variables are multiplied on the right-hand side. We may have:

$$x = abc. \quad (A4.4)$$

A4 Appendix: Decomposition with Differencing

The total derivative of (A4.4) is:

$$dx = da.bc + a.db.c + ab.dc. \tag{A4.5}$$

If we seek to represent the proportional rate of change of x, then we simply divide both sides of (A4.5) by x. This gives:

$$\frac{dx}{x} = \frac{da.bc}{x} + \frac{a.db.c}{x} + \frac{ab.dc}{x}$$

$$= \frac{da.bc}{abc} + \frac{a.db.c}{abc} + \frac{ab.dc}{abc}. \tag{A4.6}$$

Cancellation in (A4.6) gives:

$$\frac{dx}{x} = \frac{da}{a} + \frac{db}{b} + \frac{dc}{c}. \tag{A4.7}$$

Here (A4.7) is of the same form as (4.3) derived by the logarithmic approach.

It is also worth noting that this approach would work with the sum of two products. For example, suppose the relationship to be explored were:

$$x = ab + rs. \tag{A4.8}$$

The total derivative of (A4.8) is:

$$dx = da.b + a.db + dr.s + r.ds. \tag{A4.9}$$

Again, if we wish to explore the proportional rate of change of x, we divide through both sides of (A4.9) by x. This gives:

$$\frac{dx}{x} = \frac{da.b}{x} + \frac{a.db}{x} + \frac{dr.s}{x} + \frac{r.ds}{x}. \tag{A4.10}$$

Unlike equation (A4.6), equation (A4.10) cannot be further simplified in any useful way. We see, therefore, that the approach to decomposition using total derivatives is more general than that using logarithmic differentiation.

In general, any function can be decomposed into components, one component being associated with the differential of each variable. For example, in the total derivative above, the term $a.db$ is associated with the differential of b, db.

Now the differential of a variable represents a 'small change' in that variable. Thus our decomposition procedure states that the change in, say, CO_2 emission, is associated with terms involving the changes in the variables influencing CO_2 emission.

A4.3 Approximating the Differential with the Difference

Now our observations are in discrete time. Typically, information to allow the above decomposition would be collected on an annual basis. So the differentials must be approximated by differences. So equation (A4.5) in difference form becomes:

$$\Delta x = \Delta a.bc + a.\Delta b.c + ab.\Delta c. \tag{A4.11}$$

However, (A4.11) is not sufficiently precise to allow us to evaluate the elements of this difference without ambiguity, or without a remainder term. We first examine the problem of ambiguity.

A4.4 Resolving Ambiguity: Forward, Backward and Central Differences

When we use differentials, we may have an expression to evaluate such as: $b.da$.

Here the expression needs to be evaluated at a single *moment* of time, so finding the appropriate value for b presents no difficulties in principle.

The corresponding expression in difference form is: $b.\Delta a$. Now a difference refers to data which are only available at discrete moments in time. So the difference expression needs to be evaluated over a *period* of time. For example, if a and b are evaluated at times t and $t+1$, then clearly we can use:

$$\Delta a \equiv a(t+1) - a(t).$$

However, it is not clear how we should define b over the period t to $t+1$. There are three obvious possibilities:

1. $b \equiv b(t)$.

We refer to this as the 'forward difference assumption'.

2. $b \equiv b(t+1)$.

We refer to this as the 'backward difference assumption'.

3. $b \equiv [b(t) + b(t+1)]/2$.

We refer to this as the 'central difference assumption'.

That is, we can choose to weight the change in the variable a by either:

1. the *initial* value of b for that period, *or*

2. the *final* value of b for that period, *or*

3. the *mean* value of b for that period.

We choose to use the central difference assumption (i.e. weighting with the mean), for two reasons.

First, this approach gives consideration to all of the information available on variable b in each period.

Second, as we shall see, the central difference assumption makes the *exact* determination of the elements of the decomposition very straightforward.

As we shall be using central differences henceforth, with non-differenced variables evaluated at the mean of their values for any period, we need an unambiguous notation for such evaluation. We shall follow the usual practice of using a 'bar' above the variable where there is any chance of misunderstanding (e.g. '$\bar{b}$'). So we would replace '$b.da$' with '$\bar{b}.\Delta a$'. Where the context makes the meaning clear, the variable will be unadorned (e.g. '$b.\Delta a$').

A4.5 Calculating the Differencing Remainder Term

We now turn to the second problem we encounter when we move from using differentials to using differences. This relates to the remainder term that occurs when the actual evaluation takes place. For example, suppose we have the following relationship:

$x = abc.$

Differentiation gives:

$dx = da.bc + a.db.c + ab.dc.$

The corresponding central difference representation is:

$$\Delta x = \Delta a.\overline{bc} + \overline{a}.\Delta b.\overline{c} + \overline{ab}.\Delta c. \qquad (A4.12)$$

Suppose the values of the variables are given as follows:

$t = 1: \quad x = 10, \quad a = 5, \quad b = 2.0, \quad c = 1.0$

$t = 2: \quad x = 12, \quad a = 4, \quad b = 2.5, \quad c = 1.2.$

So we can calculate the corresponding means and differences as:

$\Delta x = 2, \quad \Delta a = -1, \quad \Delta b = 0.5, \quad \Delta c = 0.2$

$\overline{x} = 11, \quad \overline{a} = 4.5, \quad \overline{b} = 2.25, \quad \overline{c} = 1.1.$

We can now calculate the right-hand side of equation (A4.12). We obtain:

$[2.25 \times 1.1 \times (-1)] + [4.5 \times 1.1 \times 0.5] + [4.5 \times 2.25 \times 0.2] = 2.025.$

But we know that the correct left-hand side of equation (A4.12) is given by:

$\Delta x = 2.0.$

Clearly, the right-hand side is in error by:

$$2.025 - 2 = 0.025 \qquad (i.e. \ 1.25\%).$$

Now this is not a large error, but is worthwhile to enquire why it occurs. It results simply from the approximation made in going from the exact differential expression to the difference approximation.

Thus we may conclude that our difference expression will potentially be in error, and for completeness we should represent our difference expression as:

$$\Delta x = \Delta a . \overline{bc} + \overline{a} . \Delta b . \overline{c} + \overline{ab} . \Delta c + remainder.$$

Fortunately, it is not difficult to find exact expressions for the remainder term [Proops 1988]. As the derivation is somewhat lengthy, we simply note here the following.

1. Central differencing of a product of two variables gives a zero remainder term; i.e. for:

$$x = ab.$$

the exact central difference decomposition is:

$$\Delta x = \Delta a . \overline{b} + \overline{a} . \Delta b.$$

2. Central differencing of a product of three variables always give rise to a remainder term, which is one quarter of the product of the variable differences; i.e. for:

$$x = abc.$$

the exact central difference decomposition is:

$$\Delta x = \Delta a . \overline{bc} + \overline{a} . \Delta b . \overline{c} + \overline{ab} . \Delta c + \frac{1}{4} \Delta a \Delta b \Delta c.$$

3. Central differencing of a product of four variables always gives rise to a remainder term, itself involving four terms. The full decomposition of:

$$x = abcd.$$

is given by:

$$\Delta x = \Delta a.\overline{bcd} + \overline{a}.\Delta b.\overline{cd} + \overline{ab}.\Delta c.\overline{d} + \overline{abc}.\Delta d$$

$$+ \frac{1}{4}(\overline{a}.\Delta b \Delta c \Delta d + \Delta a.\overline{b}.\Delta c \Delta d + \Delta a \Delta b.\overline{c}.\Delta d + \Delta a \Delta b \Delta c.\overline{d}).$$

A4.6 Differencing Error-Prone Empirical Data

From the time-series data we have already discussed in earlier sections, evaluating the components of a difference equation is straightforward. However, it should be recalled that a difference equation is the discrete time analogue of a derivative, and it is well known that the derivative of a function which is not smooth can be very 'rough' indeed. Now our empirical data suffers from the problems of any economic data, in that it is not exact (i.e. it contains 'errors'), though we are confident that in general trend and shape it reflects reality.

Let us therefore consider the effect of differencing data between adjacent observations, both with and without error terms. Suppose we have data satisfying the following relationship.

$$x_t = \alpha + \beta t. \tag{A4.13}$$

Then the difference between successive terms of x_t in (A4.13) is given by:

$$\Delta x_t = x_{t+1} - x_t$$

$$= [\alpha + \beta(t+1)] - [\alpha + \beta t]$$

$$= \beta.$$

Now suppose that the relationship is as in equation (A4.13), except that associated with each observation is an error term, ε_t. So we have:

$$x_t = \alpha + \beta t + \varepsilon_t. \tag{A4.14}$$

62 A4 Appendix: Decomposition with Differencing

If we now take the difference between successive values of x_t in (A4.14), we get:

$$\Delta x_t = x_{t+1} - x_t$$

$$= [\alpha + \beta(t+1) + \varepsilon_{t+1}] - [\alpha + \beta t + \varepsilon_t]$$

$$= \beta + (\varepsilon_{t+1} - \varepsilon_t). \tag{A4.15}$$

Now we make the assumption that the error terms are independently derived from a common distribution, which has a zero mean and a non-zero variance, σ^2. So the term $(\varepsilon_{t+1} - \varepsilon_t)$ will have a zero mean and a variance of $2\sigma^2$. So the expected value of Δx_t is β, and it will have a relatively large variance compared with the original data series, x_t.

To try to overcome this problem, we can choose to take our differences between non-adjacent values of x. For example, we might choose to difference values of x which are eight periods apart; i.e. we define:

$$\Delta_8 x_{t+4} = \frac{x_{t+8} - x_t}{8}. \tag{A4.16}$$

On the left-hand side, we write Δ_8 to indicate that we are differencing over eight periods (e.g 1 to 9, 2 to 10, etc.). We use x_{t+4} to indicate the central period of the differencing range, where we would usually present the value of the difference in a graphical representation.

If we now consider an eight-period differencing when x_t is linear in t, with the error term, ε_t, as in equation (A4.16) above, then we get:

$$\Delta_8 x_{t+4} = \frac{x_{t+8} - x_t}{8}$$

$$= \frac{[\alpha + \beta(t+8) + \varepsilon_{t+8}] - [\alpha + \beta t + \varepsilon_t]}{8} \tag{A4.17}$$

$$= \beta + \frac{\varepsilon_{t+8} - \varepsilon_t}{8}.$$

A4.6 Differencing Error-Prone Empirical Data 63

Comparing equation (A4.17) with equation (A4.15), we see that the effect of the error term when an eight period difference is used is an eight-fold reduction in the effect of the error term, compared with when adjacent periods are differenced. That is, the variance of $\Delta_8 x_{t+4}$ is only one eighth of that of Δx_t. This means that the effect of using differencing between non-adjacent values achieves a considerable 'smoothing' of the resulting differenced series.

Of course, this smoothing has a price. First, period to period changes which are non-linear, but *non-random*, will be spuriously reduced. Second, the number of observations of the differenced data will be reduced. For example, using eight-period differencing means that differenced observations on the first four and last four periods are lost. However, for our purposes we consider the smoothing achieved to be worth the price of loss of some non-linear detail, and the loss of observations.

Finally, we should note that when we use an n-period difference, we also need to define an n-period mean. For example, we might difference:

$$x = ab.$$

to give:

$$\Delta_8 x_{t+4} = \Delta_8 a . \bar{b} + \bar{a} . \Delta_8 b.$$

Here we define $\bar{a}$ and $\bar{b}$ as follows:

$$\bar{a} = \frac{a_t + a_{t+8}}{2}, \qquad \bar{b} = \frac{b_t + b_{t+8}}{2}.$$

5 CO_2 Emissions by Germany and the UK

5.1 Introduction

A principal aim of this study is the establishment of policy scenarios concerning CO_2 emissions in Europe, particularly in Germany and the UK. To establish such scenarios for the future we need first to understand how CO_2 emissions have changed in the past.

In this chapter we collate and analyse time-series data on German and UK CO_2 emissions. Sections 5.2-5.6 examine CO_2 emissions by Germany since 1954 and the UK since 1950, in terms of total energy supply, GDP, types of fuel used, and industrial production structure. Sections 5.7 and 5.8 decompose the rates of changes in CO_2 emissions, for the German and UK data. (This technique was derived in Chapter 4.) Conclusions are drawn in Section 5.9.

In Germany and the UK, as in most developed countries, the principal source of CO_2 emission is the burning of fossil fuels, for the provision of economically useful energy. An appropriate place to begin our discussion is therefore the use of fossil fuels by these two countries.

5.2 Energy Use, CO_2 Emission and GDP for Germany and the UK

In seeking to understand fossil fuel use in Germany and the UK, and its role in affecting CO_2 emissions, we shall be particularly interested to explore the following:

1. The total energy use by these economies, and the mix of fossil and non-fossil fuels supplying this energy.

2. The relationship between total energy use and economic activity, as measured by Gross Domestic Product (GDP).

3. The quantity of the resulting CO_2 emission.

5.2.1 Data Sources

In Chapter 3 we discussed the relationships we found between energy use, CO_2 emission and GDP, at the level of the World, the USA and the EC. In Chapter 4 we used the technique of differencing to seek to understand the rates of change of CO_2 emissions, for the World, the USA and the EC, in terms of some economic variables. In this chapter we establish the pertinent data for Germany and the UK, and also use the differencing approach to decompose the observed rates of change of CO_2 emissions into components which we can interpret.

The data on fuel use were taken from the serial publications *Statistisches Bundesamt* [1989] for Germany, and *Energy Statistics* [Department of Energy 1981, 1989] for the UK.

In Figures 5.1 and 5.2 we show the total energy use by Germany, 1954-88, and the UK, 1950-88, and the corresponding CO_2 emissions.

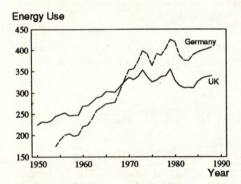 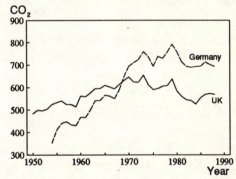

Figure 5.1. Energy use by Germany and the UK (M tce).

Figure 5.2. CO_2 emissions by Germany and the UK (M tonnes).

It is clear that both economies show a generally upward trend in energy use and are highly correlated. The UK CO_2 emission rises to a peak in 1972, and falls subsequently, while the German peak is in 1980. It is probably no coincidence that these peaks occur just around the rapid and substantial fuel price rises of 1973 and 1979. It is also clear that after 1967, energy use and CO_2 emission by Germany is considerably higher than for the UK. This can be attributed, at least in part, to the larger population of Germany than the UK. To remove the influence of population, in Figures 5.3 and 5.4 we show the *per capita* energy use and CO_2 emission for the two countries.

We see that in Figures 5.3 and 5.4 there is less difference between the two time profiles than in the case of Figures 5.1 and 5.2. We also note that there is still considerable variation in energy use and CO_2 emission over time, for both economies. The more rapid growth in both energy use and CO_2 emission by Germany can be attributed to Germany's more rapid rate of economic growth during this period.

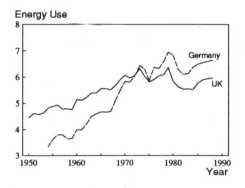 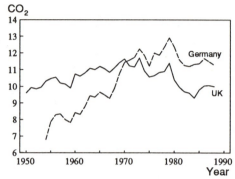

Figure 5.3. Per capita energy use by Germany and the UK (K tce per capita).

Figure 5.4. Per capita CO_2 emissions by Germany and the UK (K tonnes per capita).

5.2.2 The CO_2/Energy Ratio and the Fuel Mix

In Figures 5.3 and 5.4 we can also recognise that the profiles of energy use and CO_2 emission by Germany seem to bear a somewhat different relationship to that for the UK. This might be attributable to different fuel mixes being used in the two countries. This difference in fuel mixes would, because of the difference in carbon/energy ratios for the various fuels, give rise to different overall CO_2/energy ratios. We can see that these ratios have, indeed, been different by referring to Figure 5.5.

We note that for both Germany and the UK there is a distinct downward trend in the CO_2/energy ratio. This suggests that not-dissimilar fuel substitutions are taking place in both countries. These trends mirror the downward trend in the EC as a whole.

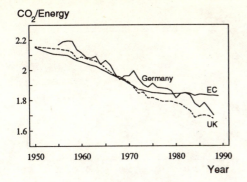

Figure 5.5. CO_2/energy for Germany, the UK and the EC (tonnes/tce).

5.3 Energy Use and GDP

It has often been suggested [Allen 1976; Adelman 1980; Proops 1984] that energy use varies with GDP. To see whether this assertion is valid for Germany and the UK during the period in question, in Figures 5.6 and 5.7 we plot energy use against GDP for the two countries (UK 1950-88, Germany 1954-1988). For comparison, we also plot CO_2 emission against GDP for Germany and the UK, in Figures 5.8 and 5.9.

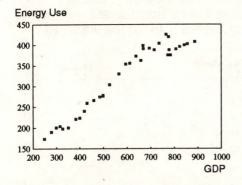
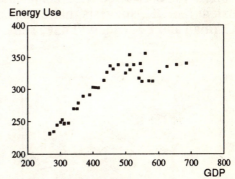

Figure 5.6. Energy use v. GDP: Germany (1954-88) (M tce; B ecu).

Figure 5.7. Energy use v. GDP: UK (1950-88) (M tce; B ecu).

We see that there is a reasonable correlation between energy use and GDP (R^2 of 0.93 and 0.77 for Germany and the UK respectively.) The relationship between CO_2 emissions and GDP is much less clear (R^2 of 0.83 and 0.18 for Germany and the UK respectively.) However, these figures suggest that there is not a *constant* ratio of energy use to GDP. Even less does it seem that there is a constant CO_2/GDP ratio.

5 CO_2 Emissions by Germany and the UK

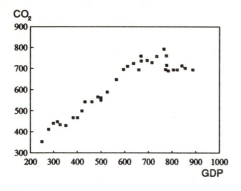

Figure 5.8. CO_2 emission v. GDP: Germany (1954-88) (M tonnes; B ecu).

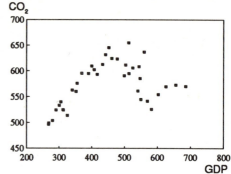

Figure 5.9. CO_2 emission v. GDP: UK (1950-88) (M tonnes; B ecu).

To test further the relationships between energy use and GDP, and CO_2 emission and GDP, in Figures 5.10 and 5.11 we plot the energy/GDP and CO_2/GDP ratios for the two countries, and for the EC as a whole. The consistent way the curves for Germany lie below those for the UK may reflect differing (but converging) industrial production and demand structures. Alternatively, they may represent a consistent (but reducing) undervaluation of the Deutschmark against the Pound Sterling. The close correlation between the EC curves and those for Germany, and to a lesser extent the UK, suggests that conclusions drawn from an analysis of the German and UK data will have some relevance also for the whole EC.

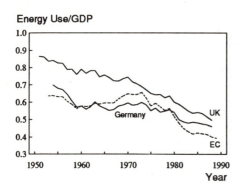

Figure 5.10. Energy use/GDP for Germany, the UK and the EC (tce/K ecu).

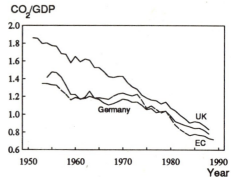

Figure 5.11. CO_2 emission/GDP for Germany, the UK and the EC (tonnes/K ecu).

We note that for both countries, and for the EC, the energy/GDP and CO_2/GDP ratios have decreasing trends over time. This may reflect a mixture of factors, such as the following:

1. *The distinction between fuel use by households and in production.*

    Changing national wealth may have a positive effect on both household and production energy demand. However, the sizes of the effects may differ. (We return to a discussion of this issue in Section 5.5, below.)

2. *The mix of industries generating national output.*

    For example, the proportion of GDP generated by iron and steel manufacture has fallen in both Germany and the UK, while the proportion of GDP generated by the services sector has increased. As it takes much more (direct) energy to produce one unit of value of steel than one unit of value of services, this change in industrial structure will cause the energy/GDP ratio to fall. (We discuss this issue in more detail in Section 5.6, below.)

3. *Changing technologies of production.*

    New technologies tend to be adopted if they are more 'economically efficient'. One type of improvement in efficiency that we observe historically is in the use of fuels. Thus one might expect technical progress will also contribute to a reduction of the energy/GDP ratio over time.

4. *Changing fuel mixes.*

    This will be influenced partly by the available technology, and partly by relative fuel prices. For example, in the UK there has been a steady shift away from coal and towards gas. This can be attributed to the discoveries of natural gas under the North Sea, which is much cheaper to extract than coal. This led to a particularly rapid substitution of gas for coal in domestic heating.

To explore in more detail the impact of various changes of fuel use, industrial structure, technical change, and fuel mix, we need to take a more disaggregated approach to fuel use. We begin by examining the breakdown of overall energy use, and CO_2 emission, for Germany and the UK, between the various fuel types.

5.4 Energy Use and CO_2 Emission by Fuel Types

In Figures 5.12 and 5.13 we show the proportional distribution of energy use by fuel type, for Germany and the UK between 1950 and 1988. This is shown as a cumulation of the three categories of fossil fuels (coal, oil and gas), and non-fossil fuel energy (hydro and nuclear power, marked 'Other'). Figures 5.14 and 5.15 show the breakdowns of the corresponding CO_2 emissions attributable to fossil fuel burning in the two countries.

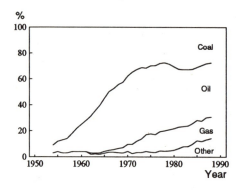

Figure 5.12. Energy use by fuel type (%): Germany.

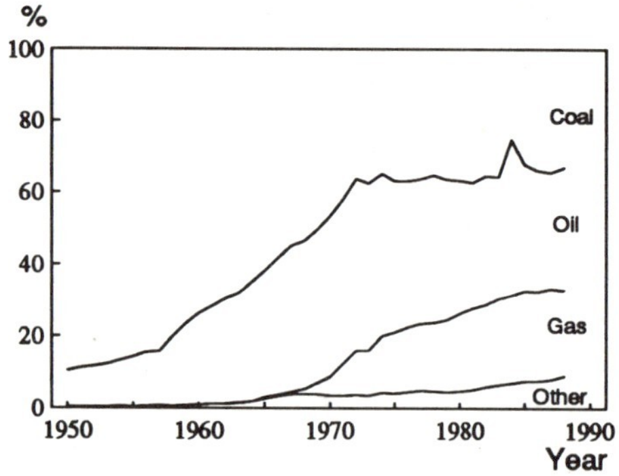

Figure 5.13. Energy use by fuel type (%): UK.

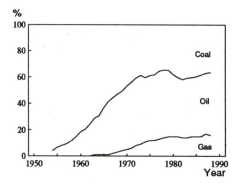
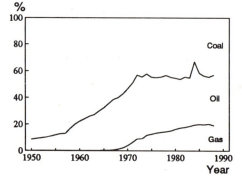

Figure 5.14. CO_2 emission by fuel type (%): Germany.

Figure 5.15. CO_2 emission by fuel type (%): UK.

We see there has been a long-run substitution of coal by oil and gas, in both Germany and the UK. Thus the proportion of CO_2 emissions attributable to oil and gas burning has also risen. From Figures 5.14 and 5.15 it is also clear that the bulk of CO_2 emission in both Germany and the UK is from burning coal and oil.

We also note the 'spike' in oil use for the UK in the mid 1980s. This reflects the 1984-5 miners' strike.

5.5 Energy Use and CO_2 Emission by Households and in Production

We noted above that we can distinguish between the energy used by households directly, and that used in production. We should recognise that, as most domestic production activity is to satisfy the needs of final consumers, the bulk of production energy use can also be attributed to household demand. (This attribution of both household *and* production energy use, and thence CO_2 emission, to 'final demand' by households is a central theme of input-output analysis, which we use extensively in later chapters of this study.)

In distributing energy use between households and industry, one needs to take care that secondary fuels, particularly electricity, are dealt with appropriately. Thus the coal, oil and gas burnt in electricity generation should *not* count as contributing only to CO_2 emission from the production sector, even though the electricity generation is clearly part of production activity. Instead, CO_2 emission by electricity generation needs to be distributed between production activity *and* households. This distribution of the CO_2 should be *pro rata* with the electricity use by these two areas.

Treating secondary fuels in this way, the historical division (1950-88) of total energy use between households and industry for Germany and the UK is shown in Figures 5.16 and 5.17 respectively.

For Germany, production energy use accounts for 63-75% of total energy use in 1954-88. It is interesting to note that energy use by households has grown substantially over this period, especially in the period 1960-70. We see that UK production energy use accounts for between 75% and 80% of total UK energy use during 1950-88. It indicates no secular trend, but it does exhibit fluctuations. On the other hand, UK household energy use (20%-25% of the total) shows a clear secular trend, with much less fluctuation.

The corresponding emissions of CO_2 by households and industry are shown in Figures 5.18 and 5.19.

72 5 CO_2 Emissions by Germany and the UK

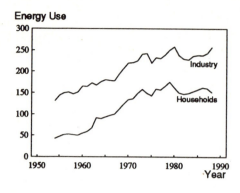

Figure 5.16. Production and household energy use: Germany (M tce).

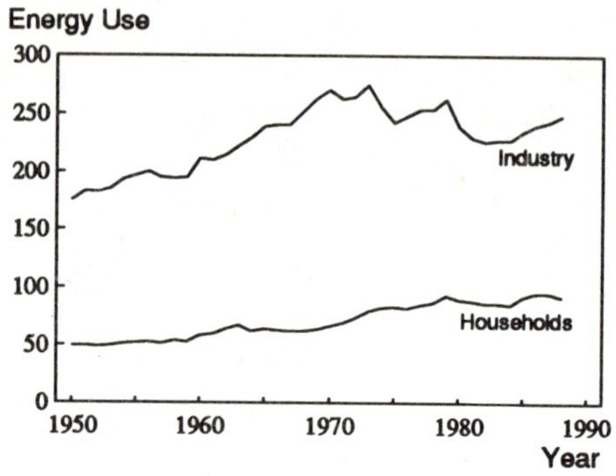

Figure 5.17. Production and household energy use: UK (M tce).

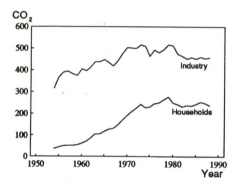
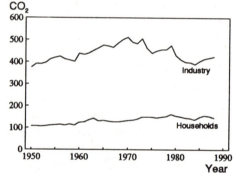

Figure 5.18. Production and household CO_2 emissions: Germany (M tonnes).

Figure 5.19. Production and household CO_2 emissions: UK (M tonnes).

We see the patterns are very similar to those for energy use; however, production CO_2 emissions clearly fall after 1969 in both Germany and the UK, while household CO_2 emissions continue to rise in both countries until 1979, especially in Germany.

5.6 Production CO_2 Emission by Sector

Having examined the way household and production CO_2 emissions have evolved, it is now useful to examine production CO_2 emission in more detail. Unfortunately, data for this exercise is not easily available for Germany, so the results below are for the UK only. We have disaggregated UK production activity into the following nine sectors:

1. Agriculture.   4. Transport.   7. Food.
2. Services.   5. Paper and Printing. 8. Chemicals.
3. Construction, etc. 6. Textiles.   9. Other Industry.

These sectors have been chosen because data are available for the UK at this disaggregation. Unfortunately, the large Sector 9 (Other Industry) contains a mix of important activities, such as Iron and Steel, and Engineering, but the data used does not allow further disaggregation.

The cumulative percentages of production CO_2 emissions by these industrial sectors is shown in Figures 5.20 and 5.21, for the UK. For clarity of presentation the nine sectors are presented in two cumulative diagrams. The sector marked 'Industry' in Figure 5.20 is then further disaggregated in Figure 5.21.

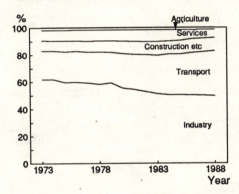

Figure 5.20. Sectoral production CO_2 emissions (%): UK (1).

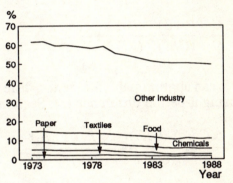

Figure 5.21. Sectoral production CO_2 emissions (%): UK (2).

We see that the proportion of CO_2 emissions from UK Industry has declined quite markedly since 1973, while that from Transport has increased quite rapidly. The proportion of Services CO_2 emission increased and then decreased. The remaining sectors have had small and relatively constant proportions of total production CO_2 emissions.

5.7 Decomposition of German and UK CO_2 Emission Changes

We can now apply the differencing decomposition technique to CO_2 emission derived in Appendix A4. We have applied the technique to decomposing the influences on CO_2 emissions by the World, the USA and the EC. We now apply the technique to Germany and the UK. We recall that we use the basic identity:

$$C \equiv \left(\frac{C}{E}\right)\left(\frac{E}{Y}\right)Y. \qquad (5.1)$$

In (5.1) we have used C for CO_2 emission in one year, E for energy use in that year, and Y for the corresponding GDP.

It will be recalled that we can logarithmically differentiate (5.1), to give:

$$\frac{\dot{C}}{C} = \frac{(\dot{C/E})}{(C/E)} + \frac{(\dot{E/Y})}{(E/Y)} + \frac{\dot{Y}}{Y}. \qquad (5.2)$$

We can approximate (5.2) with a difference expression. In this case we use an eight-period difference (cf. Section A4.6):

$$\frac{\Delta_8 C}{C} \approx \frac{\Delta_8(C/E)}{(C/E)} + \frac{\Delta_8(E/Y)}{(E/Y)} + \frac{\Delta_8 Y}{Y}. \qquad (5.3)$$

We also recall that the eight-year proportional rates of change in (5.3) can be re-expressed as annual proportional rates of change (cf. Section A4.6). It is such annualised rates that we present in the rest of this chapter. For the UK, the data range is 1950-88. As we use an eight-period difference, the range for the differenced data is 1954-1984, giving thirty-one annual observations. The German data range is 1954-88, so the range for the differenced data is 1958-1984, with twenty-seven observations.

5.7.1 Decomposition of Total CO_2 Emissions Changes

In Figures 5.22 and 5.23 the annualised eight-period proportional rates of change of CO_2 emission by the German and the UK economies are shown as the bold lines, and marked 'C'. Also shown are the three main components of these rates of change, as discussed above. These are indicated by 'C/E', 'E/Y' and 'Y'. Where the crossing of the lines can cause confusion, one or more of the lines is shown marked as dashed rather than solid. (It should be noted that the remainder term, discussed in Appendix A4, is consistently very small in the analyses that follow, so for clarity of presentation it has been omitted.)

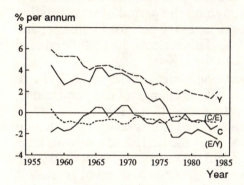

Figure 5.22. Decomposition of total CO_2 emission changes: Germany.

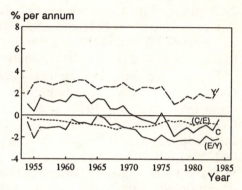

Figure 5.23. Decomposition of total CO_2 emission changes: UK.

We consider the German and UK data in Figures 5.22 and 5.23. Looking at the total CO_2 emissions (the 'C' line), we see this started positive in both countries, with a value of around 4% p.a. in Germany and 1.5% p.a. in the UK (i.e. these were the rates of increase of CO_2 emissions in the two countries). It fell to zero by 1976 in Germany, and by 1971 in the UK. It continued to fall to around -1.5% p.a. in Germany and -1% p.a. in the UK. This is very encouraging, as a sustained rate of fall in CO_2 emissions of 1% p.a. would allow the attainment of the Toronto target.

We can now explore the three elements which compose the proportional change in total CO_2 emissions.

We first note that the effect of changing national income (Y) was consistently positive throughout the period of analysis, for both Germany and the UK[1]. We assume that as income increases, so does the emission of CO_2, *ceteris paribus*. For Germany, this effect is falling from 6% p.a. to 2% p.a.

---

[1] This strict positiveness of GDP growth results from the use of 8 year differencing (see equation 5.3 above). In 1982, the annual rate of growth of GDP in Germany and the UK was negative.

For the UK, this effect is consistently around 2.5% p.a. during the period of investigation. This tells us that, had there been no countervailing effects through the other two components, CO_2 emission by the UK would have grown at approximately 2.5% p.a., and at between 2% and 6% p.a. in Germany. (These figures are simply the proportional rates of growth of GDP for Germany and the UK, respectively.)

We can see why such growth in CO_2 emissions did not occur, by looking at the component involving changes of C/E. This reflects the effect of a changing fuel mix, and we see that throughout the period of investigation this component lies between about -0.4% and -1% p.a. for Germany, and between -0.3% and -1.3% p.a. for the UK. This reflects the shift from coal to oil, and then gas, in both Germany and the UK.

Finally, we examine the component concerning changes in E/Y. In Germany, this was acting to reduce the rate of change of CO_2 emissions, up to 1964 and after 1971. The fall in the proportional rate of change of CO_2 emission in the UK can be largely attributed to this component, which falls from around -0.5% p.a. to -2% p.a. over the period of study. In both countries this effect reflects a reducing need to accompany growth in GDP with growth in aggregate fuel use. This embodies two features of economic development; improving fuel use efficiency, and a shift from manufacturing towards services.

5.7.2 Decomposition of Production CO_2 Emissions Changes

In Section 5.5 we noted that total CO_2 emissions for Germany and the UK can be divided between that attributed to households, and that attributed to production activity. The changes in these two attributed quantities can also be further divided into the three components used above. This decomposition for changes in production CO_2 emissions by industry is shown in Figures 5.24 and 5.25.

The patterns here are very similar to those for total emission. The total proportional rate of change of production CO_2 emission (C) falls from being positive early in the period to being negative.

The three components for rates of change of production CO_2 emission also closely follow those exhibited for total CO_2 emission in the two countries.

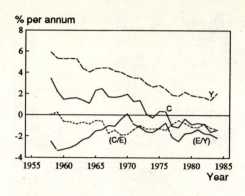
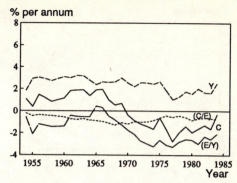

Figure 5.24. Decomposition of production CO_2 emission changes: Germany.

Figure 5.25. Decomposition of production CO_2 emission changes: UK.

5.7.3 Decomposition of Households CO_2 Emissions Changes

The same decomposition is also applied to CO_2 emissions attributable to households in Germany and the UK, in Figures 5.26 and 5.27.
The patterns here are markedly different from those for production CO_2 emission changes, and for the economy as a whole.

For Germany, the rate of change of CO_2 emissions by households begins very high, at 12% p.a., and falls to zero by 1976. This fall is partly attributable to all three components (C/E, E/Y, Y), which all fell.

For the UK, over most of the period, the rate of change of CO_2 emissions by households is positive, though with a marked dip in the mid-sixties. Since 1977 the rate of change has been negligible.

It is interesting to note that in both Germany and the UK, by the end of the study period there is an almost zero proportional rate of change of CO_2 emissions. This is because of an almost exact balancing of a positive effect through Y, and a negative effect through E/Y.

5 CO_2 Emissions by Germany and the UK

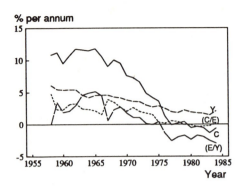

Figure 5.26. Decomposition of household CO_2 emission changes: Germany.

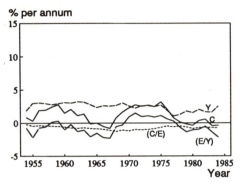

Figure 5.27. Decomposition of household CO_2 emission changes: UK.

5.7.4 Decomposition of Total CO_2 Emissions Changes

Before we move on to further interpretation and discussion of these results, it is useful to summarise the overall proportional rates of change, and their components for total, production and household CO_2 emissions. In Table 5.1 the annualised proportional rates of change are calculated, for Germany and the UK, between 1960 and 1988. That is, a twenty-eight period difference is used.

	Percent per annum				
	C	C/E	E/Y	Y	Remainder
Germany					
Total	1.64	-1.07	-1.66	4.36	0.01
Production	0.55	-1.60	-2.30	4.36	0.09
Households	5.90	1.85	-0.11	4.36	-0.02
UK					
Total	0.05	-0.86	-1.77	2.61	0.07
Production	-0.21	-0.86	-2.05	2.61	0.09
Households	0.85	-0.84	-0.90	2.61	-0.02

Table 5.1. Decomposition of CO_2 emissions changes; total, production and household: Germany and the UK (1960-88).

Regarding CO_2 emission, this increased overall in Germany, but showed almost no change in the UK. The more rapid rate of increase in Germany can be attributed mainly to the higher rate of economic growth in that country (Y).

Production CO_2 emissions were increasing in Germany overall, but decreasing in the UK. This discrepancy can also be attributed largely to the lower rate of growth of GDP in the UK.

Finally, household CO_2 emissions in Germany grew much more rapidly than in the UK. This is in part attributable to the high rate of growth of GDP. However, while in the UK the C/E ratio was falling, through substitution away from coal, in Germany it was actually rising.

5.8 Sectoral Decomposition of CO_2 Emissions Changes

We now turn to an analysis of the components of CO_2 emission changes by individual manufacturing sectors. The sectors used are those already described in Section 5.6. The relatively short time-span of the UK data (1970-88) is because of limitations on the availability of national income data on GDP contributions by the various sectors. (As already mentioned, no suitable sectoral data is available for Germany.)

The decomposition used is the same as that used in Section 5.7, except that total CO_2 emissions, total energy use and total GDP are replaced by the sectoral contributions to the total. The UK the sectoral decompositions of changes in CO_2 emissions are shown in Figures 5.28 to 5.36.

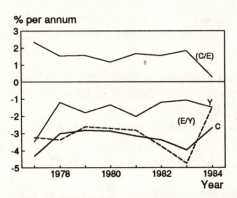

Figure 5.28. Decomposition of CO_2 emissions changes by Agriculture: UK.

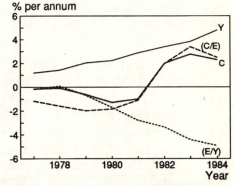

Figure 5.29. Decomposition of CO_2 emissions changes by Services: UK.

From Figure 5.28 we see that the Agriculture sector has shown a consistently negative rate of change of CO_2 emissions. It is interesting to note that, unlike the case for the whole economy, this is composed of a consistently *positive* rate of change of C/E, and a consistently *negative* rate of change of Y. The shrinkage of the output of the agricultural sector (in real terms) is outweighed by a shift towards more carbon intensive fuels. The E/Y component is also negative, indicating improving fuel efficiency by this sector.

The Services sector (Figure 5.29) indicates a slightly reducing rate of change of CO_2 emissions until 1981, with positive rates of change thereafter. As the rates of change of Y and E/Y are almost exactly opposite to each other, this behaviour is attributable almost completely to the C/E component. After 1981, Services shifted towards more carbon intensive fuels. This may reflect a move to heating with coal, subsequent to the oil price rises.

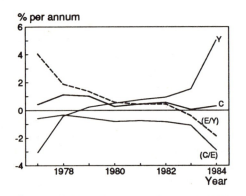
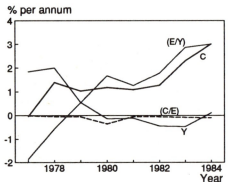

Figure 5.30. Decomposition of CO_2 emissions changes by Construction, etc.: UK.

Figure 5.31. Decomposition of CO_2 emissions changes by Transport: UK.

The Construction sector (Figure 5.30) has shown a small but consistently positive rate of change of CO_2 emissions. This is a mixture of growing rates of change of Y and reducing rates of change of E/Y and C/E.

Transport (Figure 5.31) shows a positive and increasing rate of change of CO_2 emissions, attributable almost entirely to the increase in the E/Y component. That is, the transport sector is becoming less energy efficient in the UK, and at an increasing rate.

Paper and Printing (Figure 5.32) has shown consistently negative rates of change of CO_2 emissions, despite a Y component which has been positive

5.8 Sectoral Decomposition of CO_2 Emissions Changes

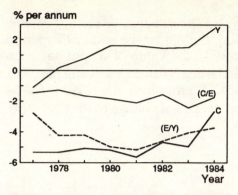

Figure 5.32. Decomposition of CO_2 emissions changes by Paper and Printing: UK.

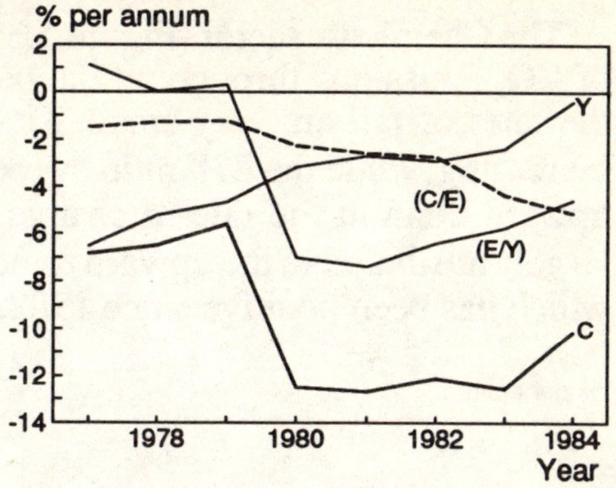

Figure 5.33. Decomposition of CO_2 emissions changes by Textiles: UK.

for most of the period. The negative C/E and strongly negative E/Y components have more than countered the effect of this sector's growth in the value of output.

Textiles (Figure 5.33) have shown a very strongly negative rate of change in CO_2 emissions throughout the period, because of shrinkage in output, increasing energy efficiency since 1979, and a shift towards less carbon intensive fuels.

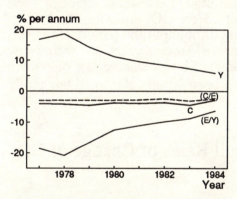
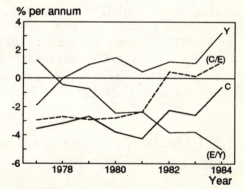

Figure 5.34. Decomposition of CO_2 emissions changes by Food: UK.

Figure 5.35. Decomposition of CO_2 emissions changes by Chemicals: UK.

The Food sector (Figure 5.34) exhibits a steadily negative rate of change of CO_2 emissions, driven almost entirely by a shift towards less carbon intensive fuels. The effects of changing sectoral output (growth) and energy efficiency (improvement) almost exactly cancel out.

The Chemicals sector (Figure 5.35) also shows negative rates of change of CO_2 emissions throughout the period, though by the end of the period the rates of fall are very small. Since 1978 the value of output has been increasing, while the E/Y ratio has been falling, at an accelerating rate. The upward trend in the rate of change of CO_2 emissions since 1981 can be largely attributed to the upward trend in the rate of change of the C/E ratio, which has been positive since 1982.

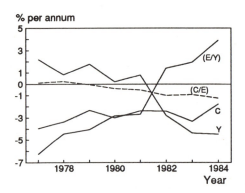

Figure 5.36. Decomposition of CO_2 emissions changes by Other Industry: UK.

Finally we look at the biggest sector (by far), Other Industry (Figure 5.36). This contains a wide range of manufacturing not elsewhere included, such as Engineering, Metal Manufacture, etc. Here CO_2 emissions have been changing at a consistently negative rate throughout the period. However, the energy efficiency has ceased to improve in recent years, and seems now to be deteriorating. This has been countered by a negative rate of growth of output since 1981. The C/E ratio has been falling, at a small but gently increasing rate, since 1979.

5.8.1 Decomposition of Overall Rates of Change of CO_2 Emission by Sectors

For completeness, in Table 5.2 we give the decomposition, by sector, of annualised rates of change of CO_2 emissions, over the entire period. These reinforce the findings discussed above.

UK (1964-88)	Percent per annum				
	C	(C/E)	(E/Y)	Y	Remainder
Agriculture	-3.00	1.34	-1.74	-2.64	0.04
Services	1.11	0.66	-2.59	3.08	-0.03
Construction	0.39	-1.24	0.55	1.08	0.00
Transport	1.70	-0.06	1.07	0.69	0.00
Paper	-2.99	-1.49	-2.64	1.12	0.03
Textiles	-7.79	-3.17	-1.88	-2.64	-0.11
Food	-2.79	-3.60	-13.55	12.91	1.45
Chemicals	-1.68	-0.69	-2.18	1.18	0.01
Other Industry	-2.29	-0.21	-2.78	0.70	0.00

Table 5.2. Decomposition of CO_2 emissions, by sector, 1970-88: UK (% p.a.).

Over the period 1964-88, only three sectors had increasing rates of CO_2 emissions; these were Services, Construction and Transport. All but Agriculture and Services showed a switch to less CO_2 intensive fuels, and only Construction and Transport seemed to become less energy efficient in their production of output. Growth in output was achieved by all sectors except Agriculture and Textiles.

5.8.2 Decomposition of Aggregate CO_2 Emission Changes

A final piece of analysis is to combine the above sectoral decompositions, to give an aggregate decomposition of UK CO_2 emissions. At first sight such aggregation would merely give again the decomposition discussed in Section 5.7.2 above. However, this decomposition of the aggregated data allows the examination of a further aspect of economic restructuring on CO_2 emissions. This aspect is the mix of industrial output.

We know that over time economies restructure themselves. For example, in most industrialised countries, over the past twenty years there has been a shift away from heavy manufacturing and towards services, measured by their proportional contributions to GDP. It is this restructuring that we can assess by reaggregating our disaggregated sectoral data. To proceed, we recall the decomposition technique applied in Section 5.8. There we used:

$$C_i \equiv \left(\frac{C_i}{E_i}\right)\left(\frac{E_i}{Y_i}\right) Y_i. \tag{5.4}$$

We extend this identity by introducing aggregate GDP, Y, where:

$$Y \equiv \sum_i Y_i. \tag{5.5}$$

Thus we write:

$$C_i \equiv \left(\frac{C_i}{E_i}\right)\left(\frac{E_i}{Y_i}\right)\left(\frac{Y_i}{Y}\right) Y. \tag{5.6}$$

We can now aggregate this relationship by summing over all sectors. Thus we obtain total production CO_2 emission as:

$$C = \sum_i C_i = \sum_i \left(\frac{C_i}{E_i}\right)\left(\frac{E_i}{Y_i}\right)\left(\frac{Y_i}{Y}\right) Y. \tag{5.7}$$

Differencing, and dividing by initial CO_2 emission, gives the following decomposition:

$$\begin{aligned}\left(\frac{\Delta C}{C}\right) &= \sum_i \Delta\left(\frac{C_i}{E_i}\right)\left(\frac{E_i}{Y_i}\right)\left(\frac{Y_i}{Y}\right)\left(\frac{Y}{C}\right) \\ &+ \sum_i \left(\frac{C_i}{E_i}\right)\Delta\left(\frac{E_i}{Y_i}\right)\left(\frac{Y_i}{Y}\right)\left(\frac{Y}{C}\right) \\ &+ \sum_i \left(\frac{C_i}{E_i}\right)\left(\frac{E_i}{Y_i}\right)\Delta\left(\frac{Y_i}{Y}\right)\left(\frac{Y}{C}\right) \\ &+ \sum_i \left(\frac{C_i}{E_i}\right)\left(\frac{E_i}{Y_i}\right)\left(\frac{Y_i}{Y}\right)\left(\frac{\Delta Y}{C}\right)\end{aligned} \tag{5.8}$$

$+ \, remainder.$

We can interpret this equation as follows:

5.8.2 Decomposition of Aggregate CO_2 Emission Changes

$$\text{Change in total } CO_2 \text{ emissions} = \text{Fuel mix effect} \\ + \text{ Energy efficiency effect} \\ + \text{ Output mix effect} \\ + \text{ Total output effect.} \quad (5.9)$$

Using the data on sectoral output and CO_2 emission, discussed in Section 5.8, we can assess the way these effects have changed over time, using equation 5.8. This decomposition is shown in Figure 5.37.

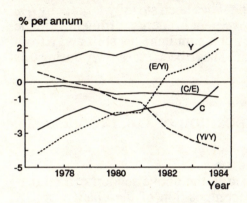

Figure 5.37. Decomposition of CO_2 emissions changes, including Output Mix Effect: UK.

We can compare this diagram with Figure 5.25, where was displayed the initial decomposition of CO_2 emissions for UK industry. Bearing in mind that Figure 5.25 is for the period 1954-84, while Figure 5.37 is for only 1975-1984, we see that three of the elements displayed in Figure 5.25 carry directly over to Figure 5.37. These are: total CO_2 emissions (C); the fuel mix effect (C/E); the total output effect (Y). From Figure 5.25 this leaves only the aggregate energy efficiency term (E/Y) which has been altered.

We note that in the original aggregate analysis, the (E/Y) factor was always negative in the period 1975-84. This suggest that the UK was producing more and more GDP with the same input of energy. However, this effect is clearly composed of two sub-effects. First, the changing efficiency of every separate industry in generating GDP from energy. Second, the mix of output that constitutes GDP. It is precisely this further decomposition which is displayed in the two elements in Figure 5.37, marked (E/Y_i) and (Y_i/Y).

In Figure 5.37, the element (E/Y_i) is what we have called the 'Energy efficiency effect' in equation (5.9). It represents the change that occurs in CO_2 emissions from UK industry, *ceteris paribus*. That is, it assumes a constant mix of output components of GDP by the various sectors. The element marked (Y_i/Y) is what we have called the 'Output mix effect' in equation (5.9). This reflects the change in CO_2 emission we have because of the changing mix of industry contributions to GDP, *ceteris paribus*.

Now from examining only Figure 5.25, one might conclude that UK is becoming more efficient at 'converting' energy into GDP. From Figure 5.37 we see that since 1981 UK industry has actually become *less* efficient at converting energy into GDP. By 1984 this component was growing at nearly 2% p.a.. Fortunately, the shift from heavy industry towards services has meant that since 1978 this component has been steadily falling, by 4% p.a. by 1984. Thus the supposed improvement in energy 'efficiency' in the UK is largely an artefact. Were there not a simultaneous shift away from heavy industry, UK CO_2 emissions would have been *increasing* during this period, rather than falling.

5.8.3 A Full Sectoral Decomposition of UK Production CO_2 Emissions Changes

The decomposition approach described in the above section can be performed at the sectoral level. This decomposition of the change of UK production CO_2 emissions in the period 1970-88 is shown in Table 5.3.

UK (1964-88)	C	(C/E)	(E/Y_i)	(Y_i/Y)	Y	Remainder
Agriculture	-0.04	0.02	-0.03	-0.07	0.03	0.00
Services	0.03	0.02	-0.06	0.03	0.05	0.00
Construction	0.03	-0.10	0.04	-0.07	0.15	0.00
Transport	0.39	-0.01	0.25	-0.29	0.45	0.00
Paper	-0.06	-0.03	-0.05	-0.02	0.04	0.00
Textiles	-0.12	-0.05	-0.03	-0.07	0.03	0.00
Food	-0.08	-0.10	-0.38	0.30	0.09	-0.01
Chemicals	-0.08	-0.03	-0.11	-0.04	0.10	0.00
Other Industry	-1.41	-0.32	-0.76	-1.43	1.07	-0.03
Total	-1.36	-0.61	-1.13	-1.66	2.00	-0.04

Table 5.3. Decomposition of production CO_2 emissions, by sector, 1970-88: UK (% p.a.).

Summing across columns 2 to 5 in Table 5.3 gives the proportional rate of change of CO_2 emission attributable to each sector. This is shown in the first column, marked 'C'. Summing down the table gives the contribution of each of the four 'effects' discussed above, in the row marked 'Total'. We see that during 1970-88, the emission of CO_2 by UK industry has been falling at an annualised rate of -1.36% p.a.. All but the 'Total output effect' have been negative, in aggregate. At the sectoral level, the only positive contributions overall are by Services, Construction and Industry. The positive Services contribution can be attributed to the growth of this sector's share in GDP, the shift towards more carbon intensive fuels (principally electricity), and the general growth in the UK economy. Construction and Transport are the only two sectors indicating a falling 'Energy efficiency effect', in the column marked (E/Y_i).

5.9 Conclusions

Both Germany and the UK have experienced a shift away from CO_2 intensive fuels in industry, although there has been a shift towards them by German households. Also, there has been a general improvement in the efficiency of 'converting' energy into GDP by industries. By the end of the period, there was also evidence of a reduction of demand for energy per unit of national income, suggesting energy may now be an inferior good.

In most UK industries, for most of the period of study, there has been a steady reduction in CO_2 emissions, despite growing output. This is attributable in part to a change in the structure of output by UK industries, and in part to a shift away from coal as a fuel. The general deterioration in overall UK energy efficiency after 1982 can be largely attributed to Construction and, very strongly, Transport.

Part III

Modelling Approach

6 A Framework for Modelling Production

6.1 Introduction

In his presidential address to the American Economic Association, Wassily Leontief [1971:6] remarked:

"An unreasonably high proportion of material and intellectual resources devoted to statistical work is now spent not on the collection of primary information but on frustrating and wasteful struggle with incongruous definitions and irreconcilable classifications".

To avoid getting entangled into such a "wasteful struggle", we consider it useful to specify carefully the theoretical basis of our study. Its cornerstone is the production structure of an economy.

The structure of the rest of this chapter is as follows. Section 6.2 discusses the nature of invention, innovation and irreversibility. In Section 6.3 we discuss activity analysis as an alternative to the more usual 'production function' approach. In particular, in Section 6.4 we relate activity analysis to input-output analysis, which we use extensively in this study. Section 6.5 compares the input-output approach to economic analysis with the applied general equilibrium approach. Finally, Section 6.6 draws some conclusions.

6.2 Invention, Innovation and Irreversibility

Our theoretical approach to analysing the greenhouse effect will be devised in such a way that it can be readily adapted to empirical research. It therefore should be suitable:

1. To examine ex post developments.

2. To predict ex ante certain features of the future evolution of the economy.

The dynamics of modern economies are characterized by the invention and innovation of techniques. Here by 'invention' the mean the making available of a new technique of production. By 'innovation' we mean the bringing into use of an invention.

Now innovation is accompanied by the building up of new capital goods and by structural change between different sectors. To consider this process in modelling, it has to be taken into account that the production of capital goods takes time, and that it has aspects of economic as well as physical irreversibilities. The latter is also true to some extent for the use of renewable resources, if their stocks fall below certain thresholds, because they may then be lost forever. Phenomena of irreversibility in the economy and the environment have to be explicitly recognised (for a more detailed analysis see Faber and Proops [1990] and Schmutzler [1991]).

To fulfil the tasks mentioned above, our theoretical approach has to be flexible and encompassing. It has to be flexible because, in the course of time, information concerning new techniques, new production factors, and new goods hitherto unknown, become available. These data have to be incorporated and the old data correspondingly adapted or erased. Our framework has to be encompassing because non-renewable and renewable resources can, in principle, be relevant for the production of *all* intermediate and final goods.

6.3 Activity Analysis vs. Production Functions

When the number of inputs in a production process is very high, it may be very cumbersome to identify the technology used by the firm when we employ the standard neoclassical production function representation of the technology. This is so because, in this case, we cannot assume that there exists substitutability between all inputs; there will also be many complementarities between the inputs. This implies that the corresponding production function has to be supplemented with other constraints, which express these various complementarities. As Malinvaud [1985:52] remarked:

> "The theory becomes very complicated if such constraints are taken into account."

The difficulties of using the concept of a production function in empirical work are well described by Dorfman, Samuelson and Solow [1958:131]:

> "The production function is a description of the technological conditions of production, and the economist takes no direct responsibility for ascertaining it. Instead he regards it as falling within the purview of the technologist or engineer. But there seems to have been a misunderstanding somewhere because the technologists do not take responsibility for the production function either. They regard the production function

as an economists' concept, and, as a matter of history, nearly all the production functions that have actually been derived are the work of economists rather than of engineers.

The engineers do not, on these grounds, stand convicted of neglect of duty. The fact is that engineers look at things somewhat differently from economists and the production function does not usually enter explicitly into an engineer's analysis."

For these reasons we tend to a very parsimonious representation of a production process. Koopmanns' [1951] notion of an activity suggests itself to model the production possibilities of individual firms and of sectors of an economy.

These considerations suggest an approach to modelling production such that it is able to deal with the following circumstances of reality:

1. Many different types of goods are produced.

2. Production takes time; capital goods must be produced before they can be used.

3. All goods are produced by *techniques*, which use some (perhaps zero) capital goods, labour and natural resources.

4. Any particular good may be produced by more than one technique.

5. Production has to be formulated such that the First Law of Thermodynamics is taken care of; i.e. that energy and matter are conserved in all transformation processes.

6. Techniques can only be run 'one-way', as determined by the Second Law of Thermodynamics; i.e. within a closed system entropy can never be decreased.

7. The set of all techniques known in one period of time is the technology of that period. Because of technical progress, depletion of resources, discovery of new resources and environmental pollution, this technology changes over time.

For theoretical reasons (see Georgescu-Roegen [1971:Ch. 9], Leontief [1971, 1982], Wodopia [1986a, 1986b, 1986c]) and also to be able to manage the great amount of data, which is necessary for empirical research, we will assume *fixed coefficient* relationships. Economists sometimes hesitate to use this assumption, because it allows no substitution between production factors during one period. But, of course, such a limitation does not exist if the operations of the model over *time* are considered, as is the purpose of our endeavour. For over time, as new, more 'efficient' techniques arise, new capital goods are accumulated, and old capital goods are not replaced as they deteriorate or are depreciated. So there is a shift from one technique to another, with consequent alterations in the capital good/output good, labour/output good and resource/output good ratios.

We purposely employed the expressions 'capital *good*' and 'output *good*' because, in this approach, the notions of homogeneous capital and output are avoided. So the aggregation which may be necessary, e.g for national accounting purposes, can only be achieved by considering the values of inputs and outputs, not simply their quantities.

The seven conditions above, and the assumption of fixed coefficients, are granted if the 'Activity Analysis' approach [Koopmans 1951] is employed. Its main element is the concept of a *Production Process*. A process is the production of one or several goods (consumption good, intermediate capital good, or waste good), using a *proportional technology*. This implies that doubling all quantities of the inputs doubles all quantities of the outputs.

Production processes can be used to define a *Technique of Production* [Bernholz and Faber 1976:349, Maier 1984:152]. A technique of production is the minimal combination of production processes, which allows the production of a consumption good from the non-produced inputs.

The *Technology* of an economy in a particular period of time is the set of all known processes in that period.

One advantage of our procedure, in contrast to a standard neoclassical production function [e.g. Burmeister and Dobell 1970:9-10], is that it yields an explicit understanding of how goods can be produced, and in particular how capital goods can be produced and accumulated, by using various techniques in the course of time. This provides an important insight into the dynamics of a modern economy, one of the main characteristics of which is technical change. For this reason we demanded above that our approach should be flexible, to allow for technical change and the resulting restructuring of the economy. For this reason it is not sufficient to restrict oneself to the substitution of production factors within a given production function, i.e. it is not sufficient to analyse only movement along isoquants of a production function. Instead we have to analyse the change of techniques (for a more detailed representation see Faber and Proops [1990:Ch. 8]).

In the following we shall show that we can employ the production process to formulate three important aspects of technical change [Faber and Proops 1990:177]:

1. *Invention* is the emergence of a new production process. This may or may not become embodied in a new technique of production.

2. *Innovation* is the process by which a new invention is brought into use through the restructuring of the entire economy. To this end new techniques are introduced and often new capital goods have to be accumulated. Old techniques and old capital goods will be substituted and finally given up.

3. *Technical Progress* is the process by which new production techniques are brought to their full potential level of efficiency. This does not involve the emergence of novelty, as the potential level of efficiency is known from consideration of the thermodynamics of the technique, etc.

In reality we observe all three kinds of technical change simultaneously in one or in different sectors. By adding and subtracting corresponding processes from the technology, we are able to model all of these changes by corresponding adaptations of the technology, in an iterative procedure over the course of time. Correspondingly, we can also deal with the emergence of new goods, production factors, resources and waste products.

Combining the activity analysis approach with capital theory allows the modelling of production and capital accumulation as a dynamic process. If the modelling takes an optimising approach, it gives rise to both quantities (from the primal solution) and to prices (from the dual solution).

6.4 Activity Analysis and Input-Output Analysis

Having shown that the activity analysis approach fulfils the requirements of being flexible and encompassing, we now turn to the question of its empirical applicability. To this end we note that the production structure of the input-output method is a special form of activity analysis. Because of the assumptions of *fixed coefficient* relationships, and of *proportional technology* (stated in the previous section), a production process of input-output analysis can be derived from a production process of activity analysis by norming the input coefficients such that the quantity of the output is equal to unity (if the process is run at unit level). (Input-output analysis is discussed more fully in the next chapter.)

In this form, however, the production processes are not yet suitable for empirical input-output analysis, even if one is only interested to carry through a study on a very disaggregated level, e.g. at the level of the firm. For even within one firm, one good may be manufactured by different production processes. For instance, assume that there are three production processes to produce a particular good, and that each process is run at unit level. Then we can aggregate the coefficients of the three processes by giving each coefficient a weight of a third (see for an illustration and comment on this procedure [Faber, Niemes and Stephan 1983:85-89]). Using this kind of procedure for the production processes of each kind of good which the firm manufactures, we obtain the production processes which are necessary for an input-output analysis of this firm.

In general, however, input-output analysis is undertaken on a sectoral level. In our study, for example, we shall use 47 sectors. Hence it is necessary to aggregate all of the production processes, of all goods, produced by all firms in a sector, to obtain one production process for this sector for the input-output analysis. Since almost every sector manufactures more than one good, this procedure implies that all the different outputs of a sector are aggregated to *one* good. (In input-output analysis this problem is dealt with by distinguishing between goods *made* by each sector (the 'Make' matrix), and the goods *purchased* by each sector (the 'Use' matrix). (For further details see Miller and Blair [1985:Ch. 5]).

From our considerations, it follows that the great amount of input-output data which is available for different years is suitable for the empirical application of the activity analysis approach, of which input-output analysis is a particular form, as discussed above. Since input-output data are also available for different countries, the activity analysis approach, in the form of input-output analysis, lends itself naturally to international studies.

6.4.1 Input-Output Analysis and National Accounting

We note that by aggregating the processes of a sector, and by aggregating sectors, one can also apply the activity analysis approach to empirical macroeconomic analysis.

Further, it is worth noting that there are close theoretical and empirical links between input-output research and national accounting (see e.g. Leontief [1973], Miller and Blair [1985], Schäfer and Stahmer [1989], Faber and Proops [1991a, 1991b]). This relationship also allows, in addition to the input-output data, the employment of the vast amount of data from national accounting for empirical research within the consistent framework of activity analysis. Using the relationship between input-output analysis and national accounting has the advantage of allowing one to work at the same time at a disaggregated microeconomic level, and at an aggregated macroeconomic level. This enables one, for example, to investigate the influence of the structure of an economy on the various components of the overall level of its Gross Domestic Product (GDP) and vice versa.

6.4.2 Activity Analysis and Dynamics

To examine the dynamics of an economy, the activity analysis approach can be readily extended, such that it can be integrated into modern capital theory (see e.g. Bliss [1975], Burmeister [1980], Faber and Proops [1990:Chs. 8, 9]). It then can be employed for the investigation of intertemporal phenomena.

To see, for instance, how technical change can be considered in an input-output framework, let us assume that in one firm an innovation of a new technique is taking place. This implies that the aggregated coefficients of production of this firm are changing over time, and correspondingly those of its sector are also changing. Since this firm is, in general, only a relatively small part of its sector, its weight is small. Hence there will also be only small changes in the coefficients of this sector. If, however, in the course of time this innovation spreads to other firms in this sector, the changes in coefficients in the sector will become larger and larger. In this way the changes of coefficients of the production processes of input-output analysis can be modelled.

6.5 Input-Output Approaches versus General Equilibrium Models

The usual justification for using an input-output approach, despite its drastic simplifications, is that it can be considered as "an adoption of the neoclassical theory of general equilibrium to the empirical study of the quantitative interdependence between interrelated economic activities" [Leontief 1966:134]. Today, fast computers allow us to use sophisticated general equilibrium models instead of the simple static input-output model.

Briefly, after Pearson [1986] for the first time addressed the CO_2 problem using an input-output approach, several applied general equilibrium models have been developed, dealing with this subject (cf. Kokoski and Smith [1987], Whalley and Wigle [1990, 1991], Hazilla and Kopp [1990], Manne [1990], Manne and Richels [1990], Bergman [1991], Conrad and Schröder [1991], Stephan et al. [1991], Alfsen [1992]).

In spite of these general equilibrium models addressing CO_2 problems, we have dispensed with a full-fledged applied general equilibrium (AGE) approach for the time being. This is because, first, most AGE models are characterised by a very high degree of aggregation, and secondly, AGE models are very demanding in terms of data requirements.

General equilibrium models are usually far more aggregated than our 47-sector approach. Therefore, they may not trace all interindustry transactions that have an impact upon CO_2 emissions.

The high degree of aggregation in most AGE models reflects the lack of adequate data. All comprehensive AGE models have very serious data availability problems, especially with respect to modelling the demand side. Therefore, very often authors have to turn to obviously unrealistic functional forms. For example, Stephan et al. [1991] assume that preferences can be represented by a CES utility function, and Blitzer et al. [1990a, 1990b] use a Stone-Geary utility function, although with such an approach it is impossible to distinguish between the properties of the model, which are imposed a priori, and the properties of the data.

Therefore, we choose to exogenise the demand side, and to assess the impact of changes in the level and structure of demand, instead of endogenising demand based on unrealistic assumptions. Additionally, in AGE models economic agents are often assumed to behave with perfect foresight over the model's time-horizon (possibly decades!). If this assumption is incorrect, the solutions of the AGE model may be more optimistic than desirable, if realistic projections of future emissions are sought.

Thus, we use an input-output model because it allows the high degree of disaggregation necessary to capture the full set of interactions between different sectors of the economy, and it accounts for emissions reacting sensitively to changing patterns of sectoral growth.

At any rate, all AGE models are based on an input-output core, and our input-output model may be interpreted as a rudimentary general equilibrium model, such that our approach may serve as a building block of an integrated long-term AGE model, to be constructed at some future date.

6.6 Conclusions

In summary, the activity analysis approach is flexible and encompassing. It can be used for theoretical and empirical analysis on a disaggregated and aggregated level for a systematic enquiry of static and dynamic microeconomic and macroeconomic problems. It is therefore this framework that we adopt for the remainder of this study.

7 Input-Output Methods

7.1 Introduction

The principal analytical technique we use in the rest of this study is input-output analysis. Many readers will already be familiar with this technique, and for these this chapter need serve only as a concise reference. However, other readers may not know of this technique (or have forgotten what they once knew!). For these readers this chapter is designed as an introduction/refresher in input-output methods.

Input-output analysis was devised in its modern form by Wassily Leontief [1936], although its roots have ben traced back to the eighteenth century 'tableau économique' of Quesnay [1758]. Those readers wishing a fuller exposition of input-output analysis than this chapter offers are warmly recommended to consult Miller and Blair [1985], and its extensive bibliography.

This chapter has the following structure. Section 7.2 outlines the concepts of input-output analysis, using scalar (element) notation. In Section 7.3 the theory is recast in the much more condensed matrix notation. Section 7.4 derives the 'Leontief inverse matrix', and discusses its central importance in input-output theory. Section 7.5 shows how the multiple interactive effects in an input-output system can be decomposed into direct and indirect effects. Section 7.6 examines how prices arise very naturally in input-output models. In Section 7.7 the technique of constructing input-output tables with financial data is discussed. Finally, in Section 7.8 the relationship between input-output tables and national accounts is outlined.

This groundwork in input-output analysis is then applied in Chapter 8, to the particular problem of how CO_2 emissions by an economy can be studied in an input-output framework.

7.2 Inter-Industry Trading and Input-Output Tables

Input-output analysis deals with an important feature of modern economic activity: inter-industry trading. Modern manufacture uses as inputs many goods that are themselves produced by manufacturing processes. For example, a motor car uses in its construction steel (from the steel industry), glass (from the glass industry), electronics (from the electronics industry), paint (from the chemicals industry), etc.

7 Input-Output Methods

Indeed, the relationships are generally even more complex than this. To make steel requires the use of electric motors, but to make electric motors requires the use of steel. To use Sraffa's [1960] phrase, we observe "the production of commodities by means of commodities".

If we are to understand these multiple interactions within an economy, we shall need a conceptual and modelling structure with three attributes.

1. Our modelling system must be disaggregated, allowing us to represent separately the various *inputs* into each productive sector, or industry. This should also allow us to distinguish between the *outputs* from each sector.

2. We should be able to distinguish between the physical *quantity* of each type of good and its economic *value*. That is, we need to be able to find an internally consistent set of prices for the commodities being produced by the economy. This corresponds to the distinction between the *primal* and *dual* representations, so familiar from modern economic analysis [Baumol 1977:Ch. 6].

3. For preference, our modelling system should be able to be represented with a simple notation, and be amenable to computation with reasonable facility.

7.2.1 A Two Sector Input-Output Model

We begin with the simplest possible example, and postulate a two-sector model economy. Each sector produces a single good, using inputs of other produced goods, and (non-produced) labour. We suppose that the goods produced can have two alternative destinations.

1. They can go to consumers where they are consumed *(Final Demand)*.

2. They can be used as inputs by one or both manufacturing sectors *(Intermediate Demand)*.

We represent the interrelationships between the producing economic sectors in an *input-output table*. In our two sector model we call the two sectors 'Industry' and 'Agriculture'. Then our table may be as shown in Table 4.1.

We first note that the numbers in this table represent *physical quantities* of goods. The representation of such relationships in *value* terms we shall turn to in Section 7.6.1

	Agriculture	Industry	Final Demand	Total Output
Agriculture	10	14	76	100
Industry	40	16	144	200
Labour	20	30		

Table 7.1. An input-output table for a two-sector model economy.

The table is easy to interpret. We simply need to remember the convention that the inputs *to* a sector are in the corresponding *column*, while the outputs *from* a sector are in the corresponding *row*.

So looking down the first column we see that the inputs into sector 1 (Agriculture) are of three types. First, 10 units of agricultural produce are themselves an input to the agricultural sector. For example, agriculture needs inputs of seed and fodder, which derives from agriculture.

The second input is 40 units of industrial goods, such as fertiliser, sprays, etc.

The third input is 20 units of labour.

Now by looking across the first row, we see that the total output of the Agriculture sector is 100 units. This goes in part to final demand (76 units), and in part to intermediate demand (24 units). We can further detail the use of the intermediate demand. The Agriculture sector uses 10 units (already noted as an input), and the Industry sector uses 14 units.

By examining the second row and column in Table 7.1, the inputs and outputs of the Industry sector can be found.

7.2.2 Activity Analysis as the Production Assumption

An input-output table is certainly an interesting representation of economic activity, giving a disaggregated, detailed, but coherent overview of economic inter-relationships. However, for the purposes of economic *analysis* it is insufficient in itself. What is also required is some modelling *assumption* about how the inputs *to* a sector are related to the outputs *from* that sector. The assumption usually made is that also used in activity analysis, discussed in the previous chapter. That is, we assume that inputs are proportional to outputs.

To see the implications of this approach for economic modelling with input-output data, we first represent Table 7.1 in algebraic form, in Table 7.2.

	Agriculture	Industry	Final Demand	Total Output
Agriculture	X_{11}	X_{12}	Y_1	X_1
Industry	X_{21}	X_{22}	Y_2	X_2
Labour	L_1	L_2		

Table 7.2. An input-output table for a two-sector model economy, in algebraic form.

We can now follow the usual activity analysis approach [Koopmans 1951], and write the relationship between the inputs to each sector and the outputs from that sector as follows:

$$\frac{X_{11}}{a_{11}} = \frac{X_{21}}{a_{21}} = \frac{L_1}{l_1} = X_1 \tag{7.1}$$

$$\frac{X_{21}}{a_{21}} = \frac{X_{22}}{a_{22}} = \frac{L_2}{l_2} = X_2. \tag{7.2}$$

The coefficients $\{a_{ij}\}$ and $\{l_i\}$ in (7.1) and (7.2) are known as the 'technological coefficients', and for any particular model are taken to be constants. Substitution and reorganisation of (7.1) and (7.2) gives the following relationships:

$$\begin{aligned} X_{11} = a_{11}X_1; \quad & X_{12} = a_{12}X_2; \quad L_1 = l_1X_1 \\ X_{21} = a_{21}X_1; \quad & X_{22} = a_{22}X_2; \quad L_2 = l_2X_2. \end{aligned} \tag{7.3}$$

These relationships can be substituted in the algebraic representation of the input-output table in Table 7.2, to give Table 7.3.

	Agriculture	Industry	Final Demand	Total Output
Agriculture	$a_{11}X_1$	$a_{12}X_2$	Y_1	X_1
Industry	$a_{21}X_1$	$a_{22}X_2$	Y_2	X_2
Labour	l_1X_1	l_2X_2		

Table 7.3. An input-output table for a two-sector model economy, in modified algebraic form.

7.2.3 Output Structure in Equation Form

For the moment we shall concentrate on the *output* structure described by this table. That is, we shall be concerned with the *rows*. Consider row 1, representing the output of the Agriculture sector. This simply says that the sum of the first three elements (intermediate demand and final demand) equals the fourth element (total output). So we can write:

$$a_{11}X_1 + a_{12}X_2 + Y_1 = X_1. \tag{7.4}$$

The second row elements, for the Industry sector, can also be combined to form an equation:

$$a_{12}X_1 + a_{22}X_2 + Y_2 = X_2. \tag{7.5}$$

Equations (7.4) and (7.5) form a pair of simultaneous equations. That is, they can be solved if all but two elements are specified. What elements might we wish to specify? Certainly we might suppose the set of technological coefficients $\{a_{ij}\}$ is known, which leaves us to choose between specifying the set of final demands $\{Y_i\}$ and the set of total outputs $\{X_i\}$. Now, as we shall discuss below, we generally feel that final demand is something over which policy makers may hope to have some direct influence, through, e.g., taxation. Total output, on the other hand, depends on not only the size and nature of the final demand for the various sectors, but also upon the nature of inter-industry trading. We therefore generally make final demand also a 'known' for our analysis, and solve our simultaneous equations for the elements of total output $\{X_i\}$.

We can solve our pair of simultaneous equations for X_1 and X_2 by reorganising equations (7.4) and (7.5) as shown below:

$$(1-a_{11})X_1 - a_{12}X_2 = Y_1 \qquad (7.6)$$

$$-a_{21}X_1 + (1-a_{22})X_2 = Y_2. \qquad (7.7)$$

Further reorganisation of equations (7.6) and (7.7) gives the solutions for X_1 and X_2 as:

$$X_1 = \frac{(1-a_{22})Y_1 + a_{12}Y_2}{(1-a_{11})(1-a_{22}) - a_{12}a_{21}} \qquad (7.8)$$

$$X_2 = \frac{a_{21}Y_1 + (1-a_{11})Y_2}{(1-a_{11})(1-a_{22}) - a_{12}a_{21}}. \qquad (7.9)$$

Now this solution is clearly somewhat complex in its structure. Also, if we extend our analysis to an economy with more than two sectors, this algebraic manipulation will be come extremely tedious, and the nature of the solution even more opaque. To overcome this we now move to a representation of our simultaneous equations in matrix form.

7.3 Matrix Representation

We recall that we can represent simultaneous equations in matrix form. For example, suppose we have the following pair of equations:

$$b_{11}x_1 + b_{12}x_2 = y_1 \qquad (7.10)$$

$$b_{21}x_1 + b_{22}x_2 = y_2. \qquad (7.11)$$

Then in matrix notation we write these equations as:

$$\begin{pmatrix} b_{11} & b_{12} \\ b_{21} & b_{22} \end{pmatrix} \begin{pmatrix} x_1 \\ x_2 \end{pmatrix} = \begin{pmatrix} y_1 \\ y_2 \end{pmatrix}. \qquad (7.12)$$

The bracketed square array on the left is a 'matrix'. The other two bracketed columns are known as 'vectors'. For reasons that will become apparent shortly, this representation is known as 'element form'.

The rule of 'matrix multiplication' is implicit in the identification of matrix equation (7.12) with simultaneous equations (7.10) and (7.11). It is given by, e.g.:

$$\begin{pmatrix} b_{11} & b_{12} \\ b_{21} & b_{22} \end{pmatrix} \begin{pmatrix} x_1 \\ x_2 \end{pmatrix} = \begin{pmatrix} b_{11}x_1 + b_{12}x_2 \\ b_{21}x_1 + b_{22}x_2 \end{pmatrix}.$$

For compactness of notation we now make the following definitions.

$$\mathbf{B} \equiv \begin{pmatrix} b_{11} & b_{12} \\ b_{21} & b_{22} \end{pmatrix}; \qquad \mathbf{x} \equiv \begin{pmatrix} x_1 \\ x_2 \end{pmatrix}; \qquad \mathbf{y} \equiv \begin{pmatrix} y_1 \\ y_2 \end{pmatrix}.$$

This very abbreviated representation of matrices and vectors is known as 'condensed form', as opposed to the above 'element form'. We follow the normal convention for condensed form, of using emboldened upper case letters for matrices, and emboldened lower case letters for vectors. This allows us to write our system of simultaneous equations as a single matrix equation:

$$\mathbf{Bx} = \mathbf{y}. \tag{7.13}$$

7.3.1 The Unit Matrix and Matrix Inversion

Finding the solution in matrix form is now straightforward. We first define a *unit matrix* (or identity matrix) $\mathbf{I}$, defined in element form by:

$$\mathbf{I} \equiv \begin{pmatrix} 1 & 0 \\ 0 & 1 \end{pmatrix}.$$

This acts for matrix multiplication as does 'unity' (i.e. '1') for scalar multiplication. e.g.:

$\mathbf{B} = \mathbf{BI} = \mathbf{IB}.$

This also allows us to define the 'inverse' of a matrix, in analogy with the process of division for scalars. We recall that for a scalar b:

$$b^{-1} \equiv \frac{1}{b}.$$

So we can write:

$$bb^{-1} = b^{-1}b = 1.$$

We speak of b^{-1} as the 'inverse' of b. We similarly define the inverse of a matrix as follows. For a regular matrix, say **B**, we define the inverse of matrix **B** as $\mathbf{B}^{-1}$, by:

$$\mathbf{BB}^{-1} = \mathbf{B}^{-1}\mathbf{B} = \mathbf{I}.$$

We should note that, unlike the scalar case, not every matrix has an inverse. A matrix which *cannot* be inverted is known as 'singular'.

Actually finding the inverse of a matrix (if it exists) is quite a complex process, particularly for large matrices. By hand it is very time-consuming, and for large matrices quite unfeasible. However, there exist efficient algorithms for computing matrix inverses (or determining that they do not exist). These are even available in most spreadsheet packages implemented on desk-top PCs. Therefore the problem of actually determining the inverse of a matrix need not detain us in this chapter, nor indeed in the rest of this study.

7.3.2 Solving Simultaneous Equations in Matrix Form

We recall that the problem we had in Section 7.3 was:

$$\mathbf{Bx} = \mathbf{y}. \tag{7.13}$$

If we can find the inverse of **B** (i.e. if **B** is 'non-singular'), then we can multiply both sides of equation (7.13) by $\mathbf{B}^{-1}$, *on the left*.[1] This multiplication gives us:

$$\mathbf{B}^{-1}\mathbf{B}\mathbf{x} = \mathbf{B}^{-1}\mathbf{y}. \tag{7.14}$$

Now we recall that $\mathbf{B}^{-1}\mathbf{B} = \mathbf{I}$, so we get:

$$\mathbf{I}\mathbf{x} = \mathbf{B}^{-1}\mathbf{y}. \tag{7.15}$$

Now multiplication by the unit matrix, **I**, leaves any matrix or vector unchanged, so we have:

$$\mathbf{x} = \mathbf{B}^{-1}\mathbf{y}. \tag{7.16}$$

This is the solution of our simultaneous equations in matrix form. Equation (7.16) gives the vector **x** in terms of the matrix **B** and the vector **y**.

Although in the above outline of elementary matrix algebra, the examples were confined to solving two simultaneous equations in two unknowns, the technique and the matrix representation can be extended to any size of problem (e.g. 148 equations in 148 unknowns). The matrix representation is identical to the two equation case, although of course the inversion of a (148×148) matrix is a considerable task in numerical manipulation, though one made easy by powerful modern computers.

7.4 The Input-Output Model in Matrix Form

Having established the basic concepts of input-output analysis, and of matrix algebra, we can now combine these to give input-output analysis a matrix representation. We recall the basic pair of equations for the output by a two sector input-output model.

---

[1] We emphasise that the multiplication is on the left, as the order of multiplication in matrix algebra is important, unlike the scalar case; i.e. for two matrices **C** and **D**, then in general $\mathbf{CD} \neq \mathbf{DC}$

$$a_{11}X_1 + a_{12}X_2 + Y_1 = X_1 \tag{7.4}$$

$$a_{12}X_1 + a_{22}X_2 + Y_2 = X_2. \tag{7.5}$$

In the element form of matrix notation we can write these equations as:

$$\begin{pmatrix} a_{11} & a_{12} \\ a_{21} & a_{22} \end{pmatrix} \begin{pmatrix} X_1 \\ X_2 \end{pmatrix} + \begin{pmatrix} Y_1 \\ Y_2 \end{pmatrix} = \begin{pmatrix} X_1 \\ X_2 \end{pmatrix}. \tag{7.17}$$

We can further simplify this representation, with condensed form, as:

$$\mathbf{Ax + y = x}. \tag{7.18}$$

We recall that **A** is the matrix of technological coefficients, **y** is the vector of final demands, and **x** is the vector of corresponding total outputs.

To be able to solve this system for the vector of total output, we need to be able to combine the two elements in the equation which contain the vector **x**. To allow this we recall that multiplication by the unit matrix, **I**, has no effect on a matrix or vector. So we can write:

$$\mathbf{x = Ix}. \tag{7.19}$$

Substitution of (7.19) into (7.18) gives:

$$\mathbf{Ax + y = Ix}. \tag{7.20}$$

We reorganise this to give:

$$\mathbf{Ix - Ax = y}. \tag{7.21}$$

Clearly, the two elements on the left-hand side of (7.21) can be combined by factoring out the vector **x**, to give:

$$\mathbf{(I - A)x = y}. \tag{7.22}$$

It is useful at this stage to write out (7.22) in element form:

$$\begin{pmatrix} 1 - a_{11} & -a_{12} \\ -a_{21} & 1 - a_{22} \end{pmatrix} \begin{pmatrix} x_1 \\ x_2 \end{pmatrix} = \begin{pmatrix} y_1 \\ y_2 \end{pmatrix}. \tag{7.23}$$

We immediately recognise equation (7.23) as being the element matrix form of equations (7.6) and (7.7).

To solve equation (7.22) for the total output vector **x**, we need to find the inverse of the matrix $(\mathbf{I}-\mathbf{A})$. Multiplying through on the left by its inverse gives us:

$$(\mathbf{I}-\mathbf{A})^{-1}(\mathbf{I}-\mathbf{A})\mathbf{x} = (\mathbf{I}-\mathbf{A})^{-1}\mathbf{y}. \tag{7.24}$$

This simplifies to:

$$\mathbf{x} = (\mathbf{I}-\mathbf{A})^{-1}\mathbf{y}. \tag{7.25}$$

Equation (7.25) is the fundamental matrix representation of input-output analysis, and it deserves close scrutiny. We begin by naming the inverse matrix. In honour of the economist who first derived this representation, and indeed founded input-output analysis, $(\mathbf{I}-\mathbf{A})^{-1}$ is known as the 'Leontief inverse'.

7.4.1 Reconstructing the Input-Output Table

To see the power of this method, we recall the input-output table in Table 7.1. It is easy to see that this has technological coefficients $\{a_{ij}\}$ given by the matrix:

$$\mathbf{A} = \begin{pmatrix} 0.1 & 0.07 \\ 0.4 & 0.08 \end{pmatrix}. \tag{7.26}$$

From (7.26) we obtain:

$$\mathbf{I}-\mathbf{A} = \begin{pmatrix} 0.9 & -0.07 \\ -0.4 & 0.92 \end{pmatrix}. \tag{7.27}$$

From (7.27) we obtain by inversion:

$$(\mathbf{I}-\mathbf{A})^{-1} = \begin{pmatrix} 1.15 & 0.0875 \\ 0.5 & 1.125 \end{pmatrix}. \tag{7.28}$$

If we now insert the above Leontief inverse and the **y** vector (final demand) from Table 7.1 into equation (7.25) we get:

$$\begin{pmatrix} 1.15 & 0.0875 \\ 0.5 & 1.125 \end{pmatrix} \begin{pmatrix} 76 \\ 144 \end{pmatrix} = \begin{pmatrix} 100 \\ 200 \end{pmatrix}.$$

On the right-hand side we have the **x** vector (total output) which corresponds to this particular matrix of technological coefficients and this particular vector of final demands. It is, of course, identical to the vector we had originally in Table 7.1. We can even reconstruct the original input-output table, by recalling our input-output equation (7.18):

Ax + **y** = **x**.

From the given **A** matrix and **y** vector we have constructed the corresponding **x** vector. We can now construct the product **Ax**, which is the intermediate flows. i.e.:

$$\mathbf{Ax} = \begin{pmatrix} 0.1 & 0.07 \\ 0.4 & 0.08 \end{pmatrix} \begin{pmatrix} 100 \\ 200 \end{pmatrix} = \begin{pmatrix} 10+14 \\ 40+16 \end{pmatrix}.$$

So if one is given a coefficients matrix, **A**, and a final demand vector, **y**, then one can construct the total output vector, **x**, and thence the intermediate demand elements, **Ax**. Thence it is a simple matter to construct the input-output table.

To illustrate the power of this technique, let us consider an economy with the technological coefficients as given in (7.26), but a slightly different vector of final demands. Let us suppose that the final demand for Agriculture (Sector 1) increases from 76 to 100. i.e.:

$$\mathbf{y} = \begin{pmatrix} 100 \\ 144 \end{pmatrix}. \tag{7.29}$$

As the matrix of technological coefficients is unchanged, so is the Leontief inverse. Therefore, all that needs to be done is to multiply the Leontief inverse already calculated in (7.28) by the new final demand vector in (7.29). This gives:

$$\mathbf{x} = (\mathbf{I} - \mathbf{A})^{-1}\mathbf{y} = \begin{pmatrix} 1.15 & 0.0875 \\ 0.5 & 1.125 \end{pmatrix} \begin{pmatrix} 100 \\ 144 \end{pmatrix} = \begin{pmatrix} 127.6 \\ 212 \end{pmatrix}. \tag{7.30}$$

We see that the new total output vector has *both* its elements increased in size, not just the output by Sector 1. To see why this is so, we next construct the intermediate demand for this model economy. We find:

$$\mathbf{Ax} = \begin{pmatrix} 0.1 & 0.07 \\ 0.4 & 0.08 \end{pmatrix} \begin{pmatrix} 127.6 \\ 212 \end{pmatrix} = \begin{pmatrix} 12.76 + 14.84 \\ 51.04 + 16.96 \end{pmatrix}. \tag{7.31}$$

So we can now substitute (7.29), (7.30) and (7.31) into our basic input-output equation:

$$\mathbf{Ax} + \mathbf{y} = \mathbf{x}.$$

This gives:

$$\begin{pmatrix} 12.76 + 14.84 \\ 51.04 + 16.96 \end{pmatrix} + \begin{pmatrix} 100 \\ 144 \end{pmatrix} = \begin{pmatrix} 127.6 \\ 212 \end{pmatrix}. \tag{7.32}$$

This is now the bulk of the input-output table for this new final demand vector. (The only thing missing is the labour input vector, to which we shall return shortly.) It is noteworthy that by changing *one* element of the final demand vector, *all* of the elements of the intermediate demand array have been changed. This is, of course, because of the importance of inter-industry trading in this model economy.

7.5 Decomposing Direct and Indirect Effects

We can think of the impact of inter-industry trading in terms of the *direct* and *indirect* effects of the elements of final demand. For example, when we increase the final demand for Sector 1 from 76 to 100, the *direct* demand for the output from sector 1 increases by 24 units. However, to meet this extra output requirement from Sector 1, the use of intermediate goods by Sector 1 from Sectors 1 *and* 2 means that there are further indirect requirements. These *first-round* indirect effects must, themselves, be produced by the economy. Thus there will also be *second-round* indirect effects, and thence *third-round* indirect effects, etc.

7 Input-Output Methods

Can we calculate these indirect effects? A moment's reflection shows that we can.

1. The first-round indirect requirement is the production necessary to allow the production of a final demand vector, **y**. This is simply **Ay**.

2. The second-round indirect requirement is the production necessary to allow the production of the first-round indirect effect vector, **Ay**. This is $\mathbf{A(Ay) = A^2y}$.

3. By similar reasoning, the third-round indirect requirement is $\mathbf{A(A^2y) = A^3y}$.

Therefore the *total* indirect effect can be written as the following infinite sum:

$$\mathbf{Ay + A^2y + A^3y + \ldots = \sum_{i=1}^{\infty} A^i y}.$$

The *total* output from the final demand vector, **y**, is therefore the direct effect plus the indirect effect. i.e.:

$$\mathbf{x = y + Ay + A^2y + A^3y + \ldots} \quad . \tag{7.33}$$

Using the identity matrix, **I**, we can rewrite: (7.33) as :

$$\mathbf{x = (I + A + A^2 + A^3 + \ldots)y}.$$

But we already know:

$$\mathbf{x = (I - A)^{-1} y}.$$

This suggest the following relationship:

$$\mathbf{(I - A)^{-1} = I + A + A^2 + A^3 + \ldots} \quad . \tag{7.34}$$

If we recall our elementary scalar theory of infinite series, this result should not be such a surprise. We recall that for a scalar b, satisfying $-1 < b < 1$ we can write:

$$\frac{1}{1-b} = 1 + b + b^2 + b^3 + \dots \quad . \tag{7.35}$$

Our matrix expansion in (7.34) is just a generalisation of the scalar relationship in (7.35). However, we note that the domain of applicability of the scalar expansion is $-1 < b < 1$. This leads us to enquire whether the domain of the matrix expansion is similarly constrained. Indeed it is; just as the 'size' of b must be limited for the expansion to be valid, so must the 'size' of the matrix **A**. The precise condition is moderately complicated, and is summarised as the Hawkins-Simon condition [Hawkins and Simon 1949, Miller and Blair 1985:35-39]. However, as long as the technological matrix is derived from an input-output table that actually describes the operations of a real, or at least 'realistic', economy, then this condition is almost certain to be satisfied.

7.6 Prices in Input-Output Models

In the above discussion of production, concentrating on output, we have been careful to refer only to *quantities* of outputs. However, the data collected in reality generally refer to *values* of outputs (and inputs). Now we know:

Value = Quantity × Price.

We now have a good understanding of how input-output analysis treats quantities of goods. If we are to be able to encompass values with input-output methods, then clearly we shall need also an input-output representation of price formation.

7.6.1 Values and Prices in Input-Output Tables

We begin by returning to our original input-output table, this time expressed in terms of the *values* of inputs and outputs. We introduce prices for the goods produced in sector 1 (p_1) and sector 2 (p_2). We also introduce a price for labour (w). Our value input-output table then has the following form.

	Agriculture	Industry	Final Demand	Total Output
Agriculture	$p_1 X_{11}$	$p_1 X_{12}$	$p_1 Y_1$	$p_1 X_1$
Industry	$p_2 X_{21}$	$p_2 X_{22}$	$p_2 Y_2$	$p_2 X_2$
Labour	$w L_1$	$w L_2$		
Total Input	$p_1 X_1$	$p_2 X_2$		

Table 7.4. A value input-output table for a two-sector model economy, in algebraic form.

The major way in which this table differs from that involving only quantities is the introduction of a new row entitled 'Total Input'. As we now have the *values* of the inputs to each sector, we can now calculate the total value of these inputs. Of course, when the input details were represented only as *quantities*, such aggregation of inputs was impossible (the problem of 'adding apples and oranges').

The reader will also note that the table has been established so that, for each sector, the value of inputs equals the value of outputs. This is the requirement for accounting balance (no costs are unaccounted for), and also for economic equilibrium (for each sector, the total value of outputs must equal the total value of inputs).[2]

In Table 7.4 we now concentrate on the two input *columns*. With this value representation we can write two input equations, one for each sector. These are:

$$p_1 X_{11} + p_2 X_{21} + w L_1 = p_1 X_1 \qquad (7.36)$$

---

[2] This is required by the equilibrium condition that there be no excess demand for any goods.

7.6.1 Values and Prices in Input-Output Tables

$$p_1 X_{12} + p_2 X_{22} + w L_2 = p_2 X_2. \qquad (7.37)$$

We can also substitute our production relations derived from the activity analysis assumption. i.e.:

$$X_{11} = a_{11} X_1; \quad X_{12} = a_{12} X_2; \quad L_1 = l_1 X_1$$
$$X_{21} = a_{21} X_1; \quad X_{22} = a_{22} X_2; \quad L_2 = l_2 X_2. \qquad (7.3)$$

Substituting in (7.35) and (7.36) from (7.3) we get:

$$p_1 a_{11} X_1 + p_2 a_{21} X_1 + w l_1 X_1 = p_1 X_1 \qquad (7.38)$$

$$p_1 a_{12} X_2 + p_2 a_{22} X_2 + w l_2 X_2 = p_2 X_2. \qquad (7.39)$$

We can cancel X_1 throughout (7.38) and cancel X_2 throughout (7.39), to give:

$$p_1 a_{11} + p_2 a_{21} + w l_1 = p_1 \qquad (7.40)$$

$$p_1 a_{12} + p_2 a_{22} + w l_2 = p_2. \qquad (7.41)$$

We can write these two equations as a single matrix equation, in element form:

$$(p_1, p_2) \begin{pmatrix} a_{11} & a_{12} \\ a_{21} & a_{22} \end{pmatrix} + w(l_1, l_2) = (p_1, p_2). \qquad (7.42)$$

Equation (7.42) has a similar structure to (7.17), except that whereas in the previous equation the **A** matrix was multiplied on the *right* by a *column* vector, now it is multiplied on the *left* by a *row* vector. We can write (7.42) in condensed matrix notation as:

$$\mathbf{p'A} + w\mathbf{l'} = \mathbf{p'}. \qquad (7.43)$$

In condensed form this tells us that the price of produced goods depends on the use of intermediate inputs, at the production price (**p'A**), and the value of non-produced inputs ($w\mathbf{l'}$).

In (7.43) we have represented the row vectors as *transposes* of the column vectors, using the usual transpose symbol, the prime. We can now introduce the identity matrix into (7.43), rearrange and solve for **p′**:

$$\mathbf{p}' = w\mathbf{l}'(\mathbf{I} - \mathbf{A})^{-1}. \tag{7.44}$$

This equation tells us that in an input-output model, the prices of the produced goods are a linear function of the price of the non-produced good (labour), through the technological coefficients embodied in the matrix **A** and the vector **l**. We see again the central importance of the Leontief inverse matrix in input-output analysis.

The present model is rather unsatisfactory, in that it supposes there is only *one* non-produced input. In reality, there will be several such inputs; as well as labour we could include capital (produced in earlier periods), entrepreneurship, information, and, of course, natural resources. However, the extension here is trivial. We replace the vector **l** by a matrix **N**, containing the non-produced input coefficients for each sector. We similarly replace the scalar wage of labour, w, by a vector of non-produced factor prices, **r**. Thus our general input-output price equation becomes:

$$\mathbf{p}' = \mathbf{r}'\mathbf{N}(\mathbf{I} - \mathbf{A})^{-1}.$$

It is worth noting that the values of the various non-produced inputs are referred to collectively as the *value-added* components.

7.7 Constructing Input-Output Tables from Value-Based Data

The above discussion of input-output analysis has been on the basis of *given* technological coefficients. That is, we have assumed that by engineering studies of the various industries, the relationship between *quantities* of inputs had been established. In reality, such studies are rarely undertaken. Instead, input-output analysts normally rely upon financial data from industries, on the *value* of sales and purchases. This, of course, presents a problem, as unless these values can be divided by the appropriate prices, the underlying quantities cannot be observed. Generally, such prices are *not* available in the appropriate form. This is because input-output data is itself an aggregation of the activities of numerous individual firms, making and selling numerous different types of goods. For example, a

7.7 Constructing Input-Output Tables from Value-Based Data

motor car factory may produce several different models of cars, small trucks, and even car components for sale on to other car manufacturers. Defining a *unit* of quantity for such an enterprise is clearly very difficult.

However, this necessity of using value data for establishing our input-output tables need not be a significant obstacle in establishing matrices of suitable input-output technological coefficients, and thus allowing the construction of an appropriate Leontief inverse. We simply need to recall the definition of the technological coefficients matrix, **A**:

$$a_{ij} \equiv \frac{X_{ij}}{X_j}. \tag{7.45}$$

Now if we are working from the usual financial data available from firms, rather than engineering data, then what will actually be observable will not be the quantities, but the values; i.e. the quantities *times* the corresponding prices. Now X_{ij} is an intermediate output of sector i, so we associate with this a price p_i. Similarly, X_j is the total output of sector j, so we associate this with the price p_j. So we can write these two values as:

$$V_{ij} \equiv X_{ij}p_i; \qquad V_j \equiv X_j p_j. \tag{7.46}$$

In analogy with (7.45) we can define a matrix of coefficients derived from the values in (7.46), which we call **B**. We use:

$$b_{ij} \equiv \frac{V_{ij}}{V_j}. \tag{7.47}$$

We can substitute in (7.47) from (7.46), to give:

$$b_{ij} = \frac{X_{ij}p_i}{X_j p_j} = a_{ij}\left(\frac{p_i}{p_j}\right). \tag{7.48}$$

So the coefficients in the technological matrix we could derive from available financial data would be the coefficients derived from engineering data, weighted by the relative prices of the sector producing the intermediate good (i), and the sector receiving the intermediate good (j).

7.7.1 Deflating Input-Output Coefficients

When, as in this study, one wishes to compare technological matrices over time, there must be a recognition that if, as is usual, the matrices are derived from financial data, then if the *relative* prices of the produced goods alter over time, then so will the observed coefficients, *even if the underlying engineering coefficients are constant*. It is therefore necessary to make deflationary adjustments to the observed coefficients.

For example, one may wish to examine how the underlying (engineering) technological coefficients change over time, when one has observations only on financially derived input-output coefficients, $\{b_{ij}\}$. One may also have (externally derived) price indices for the outputs of the various sectors, $\{\rho_i\}$. If we suppose that our observations correspond to periods '1' and '2', then we can write:

$$\frac{b_{ij}(1)}{b_{ij}(2)} = \frac{a_{ij}(1)\left(\frac{p_i(1)}{p_j(1)}\right)}{a_{ij}(2)\left(\frac{p_i(2)}{p_j(2)}\right)} \qquad (7.49)$$

$$= \left(\frac{a_{ij}(1)}{a_{ij}(2)}\right)\left(\frac{p_i(1)}{p_i(2)}\right)\left(\frac{p_j(2)}{p_j(1)}\right).$$

Now the definition of a price index is:

$$\rho_i(t) = \frac{p_i(t)}{p_i(0)}.$$

So the ratio of the indices at two moments in time equals the ratio of the prices at those two moments; i.e.:

$$\frac{\rho_i(1)}{\rho_i(2)} = \frac{\left(\frac{p_i(1)}{p_i(0)}\right)}{\left(\frac{p_i(2)}{p_i(0)}\right)} = \frac{p_i(1)}{p_i(2)}.$$

We can therefore substitute in (7.49) for the ratios of unknown prices in terms of the ratios of known price indices, giving:

$$\frac{b_{ij}(1)}{b_{ij}(2)} = \left(\frac{a_{ij}(1)}{a_{ij}(2)}\right)\left(\frac{\rho_i(1)}{\rho_i(2)}\right)\left(\frac{\rho_j(2)}{\rho_j(1)}\right). \tag{7.50}$$

Reorganising (7.50), we can find an expression for the relative sizes of the underlying (engineering) coefficients, in terms of the financially derived coefficients and the price indices:

$$\frac{a_{ij}(1)}{a_{ij}(2)} = \left(\frac{b_{ij}(1)}{b_{ij}(2)}\right)\left(\frac{\rho_i(2)}{\rho_i(1)}\right)\left(\frac{\rho_j(1)}{\rho_j(2)}\right).$$

Later in this study we make use of several input-output matrices for different times. These have all been adjusted so that their differences reflect only changes in the $\{a_{ij}\}$ coefficients, and not changes in relative prices.

7.8 Input-Output Tables and National Accounts

Finally, we present a brief section on the relationship between input-output tables and national accounts. Simply put, input-output tables are a disaggregated version of national accounts; from a complete input-output table can be derived all of the usual national accounting aggregates. For example, consider the (financial) input-output table, in Table 7.5. We first note that as the table is financially based, it is 'balanced', in the sense that for each sector, the value of total output equals the value of total input.

	Agriculture	Industry	Final Demand	Total Output
Agriculture	20	30	50	100
Industry	40	80	80	200
Labour	40	90		130
Total Input	100	200	130	

Table 7.5. An illustrative value input-output table for a two-sector model economy.

This table is a numerical version of Table 7.4, with a slight modification. It contains two extra entries, showing the total value of final demand, and the total value added (payment to labour in this simple example). Clearly, these two entries are the same. Indeed, it is easy to see that the requirement of balance in each sector *ensures* that they are always identically equal.

In national accounting terms, these two entries have a straightforward interpretation. The total value of final demand is the value of total production by the economy, net of interindustry trading. It is therefore the *National Product*. The total value added is the total payments to the primary (non-produced) factors of production. As this is income to the owners of those factors, this is the *National Income*.

It is straightforward to extend the input-output framework to allow for the deterioration of capital goods, allowing the calculation of Net National Product. Also, final demand can be broken down into its components, of consumption by households, consumption by government, investment and exports. Similarly, imports can be included in the value added and/or final demand (as a negative element) depending on whether the imports considered are or are not competitive with domestic production. Thus the full range of national accounting aggregates can be accessed from a comprehensive set of input-output data.

8 The Analysis of CO_2 Emissions with Input-Output Methods

8.1 Introduction

Having outlined input-output methods in Chapter 7, in this chapter we apply this technique to the structural analysis of CO_2 emissions by economies[1]. In Section 8.2 we lay out the basic assumptions necessary to achieve this structural description, and derive the fundamental matrix equation attributing total CO_2 emissions to final demand. Section 8.3 applies the technique of 'decomposition by differencing', derived in Chapter 4, to the matrix equation describing total CO_2 emissions. Section 8.4 further extends the analysis, to take account of the roles of imports and exports in defining the CO_2 emissions for which a country can be said to be 'responsible'. Section 8.5 examines the sensitivity of total CO_2 emissions to changes in the model parameters. In Section 8.6 the material covered is briefly reviewed.

8.2 The Input-Output Assessment of CO_2 Emissions

We begin the input-output analysis of CO_2 emissions by an economy by recalling a distinction made in Chapter 3. There we noted that some fossil fuel use, and the corresponding CO_2 emission, could be attributed to the direct use of fuels by households, and other elements of final demand. For example, the use of coal, oil and gas for domestic heating, and of petrol and diesel for personal transport, derive *directly* from household requirements.

The remainder of fossil fuel use, with its corresponding CO_2 emission, is attributable to manufacturing activity, in its broadest sense. (Here we include the generation of electricity, whatever its final use, unlike our

---

[1] There is a growing literature on input-output modelling and environmental issues (see Miller and Blair [1985:Ch. 7]). For a recent discussion of the application of input-output methods to environmental problems, see Pearson [1989].

treatment in Chapter 5). However, as manufacturing activity has, as its final purpose, the satisfaction of final demand, this CO_2 emission is also ultimately attributable to final demand.

A further distinction we need to introduce into our analysis is that between various types of fossil fuel. We normally distinguish between solid (coal), liquid (oil) and gaseous (natural gas) forms of primary carboniferous fuels, and we know that these have different CO_2 emissions per unit mass, and per unit of energy delivered.

8.2.1 An Input-Output Model of Production CO_2 Emissions

For the moment, let us concentrate on the production use of carboniferous fuels, and enquire how we might represent this in an input-output framework. To begin with, let us suppose that, from national statistics, for a certain country we can find the fossil fuel use in production, of the three types. To distinguish these fuel types, we use the notation:

$$\text{Gas} \equiv 1; \quad \text{Coal} \equiv 2; \quad \text{Oil} \equiv 3.$$

We indicate the generic fossil fuel type by the subscript 'f', where $f \in \{1, 2, 3\}$. We represent the total fossil fuel use in production as F_f, for fuel f. This is an element of the 3-vector, $\mathbf{f}$, which contains the quantities of the three fossil fuels used by producing sector.

We can relate this vector of production fuel use to the corresponding total production emission of CO_2. For fuel f, we suppose that the amount of CO_2 emission per unit of fuel is e_f. These three elements form a 3-vector, $\mathbf{e}$. If we multiply this by the total production fuel use vector, $\mathbf{f}$, then we get the total production CO_2 emission, C_{ind}; i.e.:

$$\mathbf{e'f} = C_{ind}. \tag{8.1}$$

We next need to disaggregate fuel use, $\mathbf{f}$, by producing sectors. We suppose there are n producing sectors in the economy, and indicate the generic sector by the subscript 'i', where $i \in \{1, ..., n\}$. We write the *physical quantity* of fuel type f used by sector i as F_{if}.

8.2.1 An Input-Output Model of Production CO_2 Emissions

The next step is to associate this fossil fuel use with the level of activity in that sector. Now from Chapter 7 we know that any sector produces output to satisfy both final demand and intermediate demand, to give that sector's total output, X_i. We simply extend our activity analysis assumption to fossil fuels, and assume that the use of fossil fuels by any sector is proportional to the total output from that sector. This leads us to define constants of proportionality relating total output to fossil fuel use as:

$$c_{if} = \frac{F_{if}}{X_i}. \tag{8.2}$$

Here c_{if} is an element of a $(n \times 3)$ matrix, **C**. Each element of **C** tells us the 'intensity' with which a fuel is used by a sector. The larger is a particular c_{if}, the greater is the corresponding fuel use per unit of total output.

From the definition of **C**, we see that if we multiply its transpose on the right by the total demand vector, **x**, then we get the 3-vector of production fossil fuel use, **f**; i.e.:

$$\mathbf{C'x = f}. \tag{8.3}$$

If we multiply both sides of (8.3) on the left by **e'**, and we recall (8.1), we obtain:

$$\mathbf{e'C'x} = C_{ind}. \tag{8.4}$$

This is a description of the components of production CO_2 emissions, which could be further broken down by sector and by fossil fuel type. Indeed, the elements of **e'C** could be called 'CO_2 intensities', as they indicate how much CO_2 is generated per unit of total output by each sector. However, (8.4) is not yet in a form that is directly useful for this study. In Chapter 7 we noted that government influence is upon final demand (**y**), not total demand (**x**), so we need to substitute for **x** from our input-output equation, using the Leontief inverse. We recall:

$$\mathbf{x = (I - A)^{-1} y}. \tag{8.5}$$

Substituting for **x** from (8.5) into (8.4) we obtain:

$$\mathbf{e'C'(I-A)^{-1}y} = C_{ind}. \tag{8.6}$$

We can use (8.6) to define a new vector of CO_2 intensities, $\mathbf{e'C'(I-A)^{-1}}$. These show the *direct* plus *indirect* emissions of CO_2 emission per unit of *final* demand. This becomes particularly clear if we recall the series expansion of the Leontief inverse, derived in Section 7.5:

$$\mathbf{(I-A)^{-1} = I + A + A^2 + A^3 + ...} \tag{8.7}$$

Substituting (8.7) into (8.6) gives:

$$\mathbf{e'C'(I-A)^{-1} = e'C' + e'C'A + e'C'A^2 + e'C'A^3 + ...} \tag{8.8}$$

The first element of this expansion is identical to the CO_2 intensity corresponding to *total output*; i.e. it corresponds to the *direct* effect of output on CO_2 emissions. The second element is the intensity corresponding to the first-round indirect effects, the third element corresponds to the second-round indirect effects, etc. So the elements after the first one sum to the total *indirect* effect of output on CO_2 emissions.

8.2.2 An Input-Output Model of Final Demand CO_2 Emissions

We now move on to the direct use of fuel by final demand, for such purposes as personal transport, home heating, government activity, etc. We follow the same approach as for the analysis of production fuel use, and CO_2 emissions. Just as we posited a 3-vector of fossil fuel use by sector, $\mathbf{f}$, we now consider the 3-vector of fossil fuel use for direct final demand (by households, government and exports.); we call this $\mathbf{d}$. Using the vector $\mathbf{e}$ we can relate fossil fuel use to CO_2 emission, to give CO_2 emission attributable *directly* to the final demand for fossil fuels, C_{dem}. We have:

$$\mathbf{e'd} = C_{dem}. \tag{8.9}$$

We can relate the vector **d** to the value of final demand, given by the vector **y**. Now when considering final demand, we know that it has four main components: Household Consumption; Government Consumption; Exports; Stock Building.

Now the final demand for fossil fuels for household and government consumption clearly corresponds to CO_2 emissions. However, final demand for fossil fuels corresponding to exports and stock building clearly does not, as these fuels leave the country concerned, and are used elsewhere or not used at all. Therefore, it is necessary to consider the direct final demand for fossil fuels *excluding* exports and stock building. If we are given a final demand vector, **y**, then we can modify it to represent the *domestic* (i.e. non-export) final demand, by premultiplying it by a suitable $(n \times n)$ scaling matrix, **Z**. In fact, this need contain only three non-zero elements, one each for the coal, oil and gas sectors. The elements of **Z** are simply the ratios of domestic final demand (less stock building) to total final demand, for those three sectors. Thus we need to use the modified final demand vector **Zy**.

Returning to the use of fossil fuels attributable to this modified final demand vector, we suppose the relationship is mediated by the matrix **P**. This $(n \times 3)$ matrix will also have only three non-zero elements, one for each fuel type. So we can write:

$$\mathbf{P'Zy} = \mathbf{d}. \tag{8.10}$$

Combining (8.9) with (8.10), we obtain an equation for total CO_2 emissions attributable directly to final demand:

$$\mathbf{e'P'Zy} = C_{dem}. \tag{8.11}$$

8.2.3 Production CO_2 Emissions from Non-Fossil Fuel Sources

Although fossil fuel burning is much the biggest source of CO_2 emissions attributable to economic activity in industrialised countries, there are three other minor sources of CO_2 emission. For completeness we should include these in our analysis.

The first source of production CO_2 emission derives from the production of bricks and tiles with 'carbonaceous' clays. Such clays contain carbon, and when the bricks or tiles are fired in furnaces, this also burns. This actually reduces the fuel bill for brick firing, so that carbonaceous clay is a preferred material for brick and tile making. (In fact, this carbonaceous material is clearly a fossil fuel, but is not usually accounted as such in national fuel statistics.)

The second source is the emission of CO_2 from glass making, which uses calcium carbonate as a raw material, and involves the chemical reaction:

$$CaCO_3 \rightarrow CaO + CO_2.$$

The third source of additional production CO_2 emissions is the incineration of waste materials containing plastics, and other fossil-derived materials. We stress that the combustion of wastes derived from plant material, such as wood, need not be included, as this CO_2 production would have occurred in any case, with the natural degradation of this plant material.[2]

If we name these non-fossil fuel types of CO_2 emissions C_{nf}, then we can introduce a vector of non-fossil CO_2 direct intensities, **m**, such that:

$$\mathbf{m'x} = C_{nf}. \tag{8.12}$$

Recalling the relationship between total output and final demand, through the Leontief inverse, we can substitute (8.5) into (8.12), to give:

$$\mathbf{m'(I-A)^{-1}y} = C_{nf}. \tag{8.13}$$

---

[2] It should be noted that although the manufacture of cement is a cause of a considerable amount of CO_2 emission, the net effect of cement *use* is zero, as when the cement is 'curing', CO_2 is reabsorbed from the atmosphere. In this way cement production is akin to brewing and sustained agriculture. All of these processes 'recycle' CO_2.

8.2.4 An Equation for Total CO_2 Emissions

We now define the total CO_2 emissions we shall be considering in this study as C, given by:

$$C \equiv C_{ind} + C_{dem} + C_{nf}. \tag{8.14}$$

Substituting from (8.6), (8.11) and (8.13) into (8.14), we obtain:

$$C = \mathbf{e'C'(I-A)^{-1}y} + \mathbf{e'P'Zy} + \mathbf{m'(I-A)^{-1}y}. \tag{8.15}$$

Factorisation of (8.15) gives:

$$C = [\mathbf{e'P'Z} + (\mathbf{e'C'} + \mathbf{m'})(\mathbf{I-A})^{-1}]\mathbf{y}. \tag{8.16}$$

Equation (8.16) will form the basis for all of the analysis of CO_2 emissions in the rest of this study. It is therefore appropriate that we spend some time on its detailed interpretation.

The first point to note is that the total CO_2 emission by an economy, described by (8.16), is *all* attributed to final demand. This is a very desirable feature of this representation because, as mentioned previously, it is final demand (especially consumer behaviour and government expenditure) that policy makers can hope to influence reasonably directly.

We further note that we have distinguished between CO_2 emissions resulting from the demand by households and government for fuels, and that resulting from final demand for goods generally. So we distinguish between the CO_2 attributable to the manufacture of a motor car, and that attributable to its use, through the purchase of petrol.

We can further decompose (8.16), if we choose, using the series expansion of the Leontief inverse in (8.7). This allows us to distinguish between the CO_2 emissions by production activity attributable *directly* to final demand, and those CO_2 emissions attributable *indirectly* to final demand, through inter-industry trading.

The elements of (8.16) (i.e. C, **e**, **P**, **Z**, **C**, **A**, **y**) are available from published national statistics (at least for industrialised countries). This is a great advantage in establishing the database for analysis of the relationship between economic structure and CO_2 emissions. (This should not be taken to imply that establishing such a database is easy; it is very time consuming, and requires great care and the use of experienced judgement.)[3]

Finally, (8.16) is an excellent basis for scenario analysis. It contains factors reflecting the mix of fuel use by households and in production, and their 'efficiency' (**P**, **C**), the structure of inter-industry trading (**A**), and the structure and level of the final demand for goods (**y**). These can be varied separately or together, to give estimates of total CO_2 emissions by an economy under various assumptions on technological change in fuel use, the nature of inter-industry trading, and consumer and government behaviour. Such scenarios will form the subject of Part IV of this study.

8.3 Comparing CO_2 Emissions Over Time and Between Countries

If we are to understand how structural economic change in the future may affect CO_2 emissions, it would be useful if we had a tool to explore how historical structural economic change has given rise to changing CO_2 emissions. Fortunately, such a tool of analysis is readily available.

The reader is reminded that in Chapter 4, and Appendix A4, the technique of 'decomposition by differencing' was discussed and applied. We recall that the basis of this technique is the total derivative of a product. For example, we might take the differential of:

$$x = abc$$

to give:

$$dx = da.bc + a.db.c + ab.dc.$$

We also recall that if the differential is approximated by a central difference, using the 'bar' symbol to represent the mean value of a variable, this relationship becomes:

$$\Delta x = \Delta a.\overline{bc} + \overline{a}.\Delta b.\overline{c} + \overline{ab}.\Delta c + remainder.$$

---

[3] The matrix **m** can be found from chemical analysis, and data available from appropriate firms. As the size of the effect was found to be negligible, while it is included in the algebra, for completeness, it is ignored in the actual calculations.

8.3 Comparing CO_2 Emissions Over Time and Between Countries

We recall that in Chapter 5, where we applied the differencing technique to CO_2 emissions by Germany and the UK, we distinguished between aggregate growth of GDP (Y) and the change in GDP mix between sectors. We use the same distinction here, and implement this by recalling equation (8.16).

$$C = [e'P'Z + (e'C' + m')(I - A)^{-1}]y. \tag{8.16}$$

If we now define the GDP share of sector i by:

$$u_i \equiv \frac{y_i}{Y}.$$

This allows us to write the vector of final demands by each sector (**y**) in terms of the vector of GDP shares (**u**) and the level of GDP (Y), as:

$$y = uY. \tag{8.17}$$

Thus we can substitute (8.17) into (8.16) to give:

$$C = [e'P'Z + (e'C' + m')(I - A)^{-1}]uY. \tag{8.18}$$

We can apply the technique of decomposition by differencing to (8.18), to give:

$$\begin{aligned}\Delta C = & \; e'[\Delta P'\overline{Zu Y} + \overline{P}'\Delta Z\overline{uY} + \overline{P'Z}\Delta u\overline{Y} + \overline{P'Zu}\Delta Y] \\ & + e'[\Delta C'\overline{(I-A)^{-1}uY} + \overline{C}'\Delta(I-A)^{-1}\overline{uY} \\ & \quad + \overline{C}'\overline{(I-A)^{-1}}\Delta u\overline{Y} + \overline{C}'\overline{(I-A)^{-1}u}\Delta Y] \\ & + [\Delta m'\overline{(I-A)^{-1}uY} + \overline{m}'\Delta(I-A)^{-1}\overline{uY} \\ & \quad + \overline{m}'\overline{(I-A)^{-1}}\Delta u\overline{Y} + \overline{m}'\overline{(I-A)^{-1}u}\Delta Y] \\ & + remainder. \end{aligned} \tag{8.19}$$

In equation (8.19), the 'Δ' operator can be interpreted in two ways. First, it can be the difference between the values of the variables for consecutive input-output tables for a single country. Alternatively, it can be the difference between the values of the variables between two different countries, at given times.

As the term $\Delta(\mathbf{I}-\mathbf{A})^{-1}$ effectively involves the comparison of the **A** matrices for different years and/or the two countries involved, these matrices will, of course, need to be adjusted for relative price changes/differences between the sectors. That is, they will need to have been deflated, using the technique discussed in Chapter 7. We see that equation (8.19) falls into three parts, as follows:

1. The first line represents changes/differences in the emission of CO_2 because of changes/differences in the use of fuels by domestic consumers. These in turn depend on changes/differences in the cost of fuels to consumers ($\Delta\mathbf{P}'$), changes/differences in the proportion of final demand for fuels for final use ($\Delta\mathbf{Z}$), the changes/differences in the structure of final demand for the various fuel types ($\Delta\mathbf{u}$), and changes in the level of GDP (ΔY).

2. The second and third lines represent changes/differences in the emission of CO_2 because of changes/differences in the use of fuels in production. These depend on changes/differences in the technology of fuel use ($\Delta\mathbf{C}'$), changes/differences in the technology of the production of goods ($\Delta(\mathbf{I}-\mathbf{A})^{-1}$), changes/differences in the structure of GDP ($\Delta\mathbf{u}$), and changes in the level of GDP (ΔY).

3. The fourth and fifth lines represent the changes/differences in emission of CO_2 because of the use of materials other than fossil fuels in manufacturing. These depend on changes/differences in the technology of the uses of these materials ($\Delta\mathbf{m}'$), changes/differences in the technology of the production of goods ($\Delta(\mathbf{I}-\mathbf{A})^{-1}$), changes/differences in the structure of GDP ($\Delta\mathbf{u}$), and changes in the level of GDP (ΔY).

This decomposition will also reveal how far changing relative fuel prices have altered the relative use of fuels, over time and between countries, within each of the three categories above. This information will, in turn, be useful in assessing how far and how rapidly policy-determined relative price changes/differences of fuels are likely to be effective in reducing CO_2 emissions, and with what effects on economic growth.

8.4.1 Exports and the Attribution of CO_2 Emissions

As the components of CO_2 emissions derived from the non-fossil fuel sources are generally small, it simplifies the representation if these are combined with the production CO_2 emission elements. This gives:

$$\Delta C = \mathbf{e'}[\Delta \mathbf{P'} \overline{\mathbf{Zu}} \overline{Y} + \overline{\mathbf{P'}} \Delta \mathbf{Z} \overline{\mathbf{u}} \overline{Y} + \overline{\mathbf{P'}} \overline{\mathbf{Z}} \Delta \mathbf{u} \overline{Y} + \overline{\mathbf{P'}} \overline{\mathbf{Zu}} \Delta Y]$$

$$+ [(\mathbf{e'} \Delta \mathbf{C'} + \Delta \mathbf{m'}) \overline{(\mathbf{I} - \mathbf{A})}^{-1} \overline{\mathbf{u}} \overline{Y} + (\mathbf{e'} \overline{\mathbf{C}} + \overline{\mathbf{m'}}) \Delta (\mathbf{I} - \mathbf{A})^{-1} \overline{\mathbf{u}} \overline{Y}$$
$$\quad\quad (8.20)$$
$$+ (\mathbf{e'} \overline{\mathbf{C'}} + \overline{\mathbf{m'}}) \overline{(\mathbf{I} - \mathbf{A})}^{-1} \Delta \mathbf{u} \overline{Y} + (\mathbf{e'} \overline{\mathbf{C'}} + \overline{\mathbf{m'}}) \overline{(\mathbf{I} - \mathbf{A})}^{-1} \overline{\mathbf{u}} \Delta Y]$$

$+ remainder.$

It is this decomposition that is used for the empirical work in Part III of this study.

8.4 Imports, Exports and 'Attributable' CO_2 Emissions

The above calculations allow one to identify CO_2 emissions associated with various types of fuel use, by sector, and explore how they vary over time, and between countries. However, the CO_2 emission that takes place within any country may not be entirely its 'responsibility'. For example, if country A produces steel that it exports to country B, to allow country B to produce manufactured goods with that steel, then one should allocate responsibility for the CO_2 emission attributable to this steel production to country B, *not* country A.

In the above methodology, CO_2 emission is always associated with the final demand (directly or indirectly) for goods and services. As goods which are exports are almost certainly associated with *other* countries' final demands, the responsibility for this CO_2 emission should be allocated to those other countries. Similarly, goods imported should be included in the attribution of CO_2 responsibility to the importing country.

8.4.1 Exports and the Attribution of CO_2 Emissions

We first recall the basic I-O equation concerning the emission of CO_2 by a single country. This is:

$$C = [e'P'Z + (e'C' + m')(I-A)^{-1}]y. \qquad (8.16)$$

We can subdivide the vector of final demand (y) into two components. The first component (y^d) corresponds to domestic final demand (by households, government, and for investment). The second component (y^e) corresponds final demand for exports. So we can write:

$$y = y^d + y^e. \qquad (8.21)$$

We apply this decomposition of y to equation (8.16). This gives:

$$C = e'P'Zy^d + [(e'C' + m')(I-A)^{-1}](y^d + y^e). \qquad (8.22)$$

The reader will note that the term $e'P'Zy^e$ has been suppressed, as this would have no sensible interpretation. We can further reorganise (8.22) to give:

$$C = e'P'Zy^d + (e'C' + m')(I-A)^{-1}y^d \qquad (8.23)$$
$$+ (e'C' + m')(I-A)^{-1}y^e.$$

We can interpret equation (8.23) as follows. The first part of the right-hand side represents the CO_2 emission attributable directly to households. The second part corresponds to the CO_2 emission attributable to fossil fuel use for producing goods for domestic final demand. The third part corresponds to CO_2 emission attributable to the production of final demand for export.

The CO_2 emission attributable to exports should be subtracted from the country's total CO_2 emission, as responsibility for this emission should be attributed to the importing country.

8.4.2 Imports and the Attribution of CO_2 Emissions

Turning now to the treatment of imports by a country, these are probably in the main to meet that country's final demand, so responsibility for the CO_2 emission associated with imports should largely devolve on the importing country.

The above discussion is a little oversimplified, as it ignores the distinction between imports for use by consumers directly, and that for use in further production. That is, imports may be directly into final demand, or they may be to satisfy intermediate demand. If imports are to intermediate demand, then these imports may give rise to products which are themselves exported. These exports could, in turn, be to other countries' intermediate demand, and be re-exported, back to the original importing country.

Clearly, such feedback between regions is possible, just as feedback between sectors can take place within a region. To analyse the full set of possible inter-regional feedbacks that can occur, an extended input-output methodology is required. This methodology is presented in the next section.

8.4.3 Calculating CO_2 Emission in a 2-Region, n-Sector Economy

For simplicity, we consider a model with 2 regions. We describe the economic activity in this model with the following inter-regional trade input-output table:

	Region 1	Region 2	Final Demand		Total Demand
Region 1	X_1	H_1	y^{11}	y^{12}	x^1
Region 2	H_2	X_2	y^{21}	y^{22}	x^2

Table 8.1. An inter-regional trade input-output table.

In Table 8.1 we have used the following notation:

X_α - Matrix of intermediate flows for region α.
H_α - Matrix of import flows from region α ($\alpha = 1, 2$) to region β ($\beta = 1, 2$; $\alpha \neq \beta$).
$y^{\alpha\beta}$ - Vector of final demand from region α delivered to region β.
x^α - Vector of total demand by region α.

We make the usual assumption on technical relations. We define the (intra-region) technological coefficients matrix for region α by:

$$a_{ij}^\alpha \equiv \frac{x_{ij}^\alpha}{x_j^\alpha}. \tag{8.24}$$

We use (8.24) to define the *intra-region* technological coefficients matrix for region α as $\mathbf{A}_\alpha$.

By analogy, we can define the *inter-region* technological coefficients for region α as:

$$b_{ij}^\alpha \equiv \frac{h_{ij}^\alpha}{x_j^\beta}. \tag{8.25}$$

We use (8.25) to define the technological coefficients matrix $\mathbf{B}_\alpha$, which is the imports coefficients matrix, for imports to intermediate demand.

Turning now to final demand, we define:

$$\mathbf{y}^\alpha \equiv \mathbf{y}^{\alpha\alpha} + \mathbf{y}^{\alpha\beta}. \tag{8.26}$$

Here $\mathbf{y}^\alpha$ is the vector of final demand for goods from region α, irrespective of where the goods are consumed.

Using the structure of Table 8.1, and (8.24), (8.25) and (8.26), we can write:

$$\begin{pmatrix} \mathbf{A}_1 & \mathbf{B}_1 \\ \mathbf{B}_2 & \mathbf{A}_2 \end{pmatrix} \begin{pmatrix} \mathbf{x}^1 \\ \mathbf{x}^2 \end{pmatrix} + \begin{pmatrix} \mathbf{y}^1 \\ \mathbf{y}^2 \end{pmatrix} = \begin{pmatrix} \mathbf{x}^1 \\ \mathbf{x}^2 \end{pmatrix}. \tag{8.27}$$

We can solve (8.28) using the technique for the inversion of a partitioned matrix [Johnston 1972:90], to give:

$$\begin{pmatrix} (\mathbf{I}-\mathbf{A}_1) & -\mathbf{B}_1 \\ -\mathbf{B}_2 & (\mathbf{I}-\mathbf{A}_2) \end{pmatrix}^{-1} \begin{pmatrix} \mathbf{y}^1 \\ \mathbf{y}^2 \end{pmatrix} = \begin{pmatrix} \mathbf{x}^1 \\ \mathbf{x}^2 \end{pmatrix}. \tag{8.28}$$

As usual, we suppose we know total CO_2 production, by regions (α), given by:

$$\mathbf{C}^\alpha = \mathbf{r}^{\alpha\prime} \mathbf{x}^\alpha. \tag{8.29}$$

8.4.3 Calculating CO_2 Emission in a 2-Region, n-Sector Economy

Here, **r** is an n-vector, defined by $\mathbf{r}' \equiv \hat{\mathbf{e}}\mathbf{C}'$, where $\hat{\mathbf{e}}$ is a (3×3) matrix, with the vector **e** on the diagonal; i.e:

$$(\mathbf{r}^{1\prime}, \mathbf{r}^{2\prime})\begin{pmatrix} \mathbf{x}^1 \\ \mathbf{x}^2 \end{pmatrix} = C \quad (= C^1 + C^2). \tag{8.30}$$

So substituting for $(\mathbf{x}^1, \mathbf{x}^2)'$ from (8.28), we get:

$$(\mathbf{r}^{1\prime}, \mathbf{r}^{2\prime})\begin{pmatrix} (\mathbf{I}-\mathbf{A}_1) & -\mathbf{B}_1 \\ -\mathbf{B}_2 & (\mathbf{I}-\mathbf{A}_2) \end{pmatrix}^{-1}\begin{pmatrix} \mathbf{y}^1 \\ \mathbf{y}^2 \end{pmatrix} = C. \tag{8.31}$$

We can disaggregate (8.31) by writing final demand as:

$$\begin{pmatrix} \mathbf{y}^1 \\ \mathbf{y}^2 \end{pmatrix} \to \begin{pmatrix} \mathbf{y}^{11} & \mathbf{y}^{12} \\ \mathbf{y}^{21} & \mathbf{y}^{22} \end{pmatrix}. \tag{8.32}$$

Also we can make the disaggregation:

$$(\mathbf{r}^{1\prime}, \mathbf{r}^{2\prime}) \to \begin{pmatrix} \mathbf{r}^{1\prime} & 0 \\ 0 & \mathbf{r}^{2\prime} \end{pmatrix}. \tag{8.33}$$

Substituting (8.32) and (8.33) into (8.31) gives:

$$\begin{pmatrix} \mathbf{r}^{1\prime} & 0 \\ 0 & \mathbf{r}^{2\prime} \end{pmatrix}\begin{pmatrix} (\mathbf{I}-\mathbf{A}_1) & -\mathbf{B}_1 \\ -\mathbf{B}_2 & (\mathbf{I}-\mathbf{A}_2) \end{pmatrix}^{-1}\begin{pmatrix} \mathbf{y}^{11} & \mathbf{y}^{12} \\ \mathbf{y}^{21} & \mathbf{y}^{22} \end{pmatrix} = \begin{pmatrix} C^{11} & C^{12} \\ C^{21} & C^{22} \end{pmatrix}. \tag{8.34}$$

Here we have used: $C^1 = C^{11} + C^{12}$ and $C^2 = C^{21} + C^{22}$. Next, we write:

$$\begin{pmatrix} (\mathbf{I}-\mathbf{A}_1) & -\mathbf{B}_1 \\ -\mathbf{B}_2 & (\mathbf{I}-\mathbf{A}_2) \end{pmatrix}^{-1} \equiv \begin{pmatrix} \mathbf{M} & \mathbf{Q} \\ \mathbf{R} & \mathbf{S} \end{pmatrix}. \tag{8.35}$$

Substituting (8.35) into (8.34) gives:

136 8 The Analysis of CO_2 Emissions with Input-Output Methods

$$\begin{pmatrix} \mathbf{r}^{1\prime} & 0 \\ 0 & \mathbf{r}^{2\prime} \end{pmatrix} \begin{pmatrix} \mathbf{M} & \mathbf{Q} \\ \mathbf{R} & \mathbf{S} \end{pmatrix} \begin{pmatrix} \mathbf{y}^{11} & \mathbf{y}^{12} \\ \mathbf{y}^{21} & \mathbf{y}^{22} \end{pmatrix}$$

$$= \begin{pmatrix} \mathbf{r}^{1\prime}\mathbf{M} & \mathbf{r}^{1\prime}\mathbf{Q} \\ \mathbf{r}^{2\prime}\mathbf{R} & \mathbf{r}^{2\prime}\mathbf{S} \end{pmatrix} \begin{pmatrix} \mathbf{y}^{11} & \mathbf{y}^{12} \\ \mathbf{y}^{21} & \mathbf{y}^{22} \end{pmatrix}$$

$$= \begin{pmatrix} \mathbf{r}^{1\prime}\mathbf{M}\mathbf{y}^{11} + \mathbf{r}^{1\prime}\mathbf{Q}\mathbf{y}^{21} & \mathbf{r}^{1\prime}\mathbf{M}\mathbf{y}^{12} + \mathbf{r}^{1\prime}\mathbf{Q}\mathbf{y}^{22} \\ \mathbf{r}^{2\prime}\mathbf{R}\mathbf{y}^{11} + \mathbf{r}^{2\prime}\mathbf{S}\mathbf{y}^{21} & \mathbf{r}^{2\prime}\mathbf{R}\mathbf{y}^{12} + \mathbf{r}^{2\prime}\mathbf{S}\mathbf{y}^{22} \end{pmatrix}.$$

(8.36)

So we can write the total CO_2 emission as the sum of eight components. We associate four of these components with each region.

First, we note that the CO_2 emission in a region is given by the sum of the elements with final demand *produced* in the region. So for region 1 we have:

$$C^1 \equiv \mathbf{r}^{1\prime}\mathbf{M}\mathbf{y}^{11} + \mathbf{r}^{1\prime}\mathbf{M}\mathbf{y}^{12} + \mathbf{r}^{1\prime}\mathbf{Q}\mathbf{y}^{21} + \mathbf{r}^{1\prime}\mathbf{Q}\mathbf{y}^{22}.$$

(8.37)

We can identify the components in (8.37) as all deriving from region 1, as the region 1 technology of fuel use is involved in each (i.e. each contains $\mathbf{r}^{1\prime}$).

We distinguish this from the CO_2 emission *attributable* to each region, in the sense that it is associated with final demand *delivered* in that region (i.e. the CO_2 emission for which that region is *responsible*). The components we attribute to final demand in region 1 are:

$$G^1 \equiv \mathbf{r}^{1\prime}\mathbf{M}\mathbf{y}^{11} + \mathbf{r}^{1\prime}\mathbf{Q}\mathbf{y}^{21} + \mathbf{r}^{2\prime}\mathbf{R}\mathbf{y}^{11} + \mathbf{r}^{2\prime}\mathbf{S}\mathbf{y}^{21}.$$

(8.38)

We can identify these terms as being attributable to region 1, as the final demand element in each term has '1' as its second superscript, indicating the final demand derives from region 1.

Now we can find $\mathbf{M}, \mathbf{Q}, \mathbf{R}, \mathbf{S}$ explicitly, by using the rule for inverting partitioned matrices (see e.g. Johnston [1972:90]). They are:

$$\mathbf{M} = \mathbf{G} \qquad [\mathbf{G} \equiv ((\mathbf{I} - \mathbf{A}_1) - \mathbf{B}_1(\mathbf{I} - \mathbf{A}_2)\mathbf{B}_2)^{-1}] \qquad (8.39)$$

$$\mathbf{Q} = \mathbf{G}\mathbf{B}_1(\mathbf{I} - \mathbf{A}_2)^{-1} \qquad (8.40)$$

8.4.3 Calculating CO_2 Emission in a 2-Region, n-Sector Economy

$$R = (I - A_2)^{-1} B_2 G \tag{8.41}$$

$$S = (I - A_2)^{-1} (I + B_2 G B_1 (I - A_2)^{-1}). \tag{8.42}$$

Now if we assume 'Region 1' is 'Germany' (or 'UK'), and 'Region 2' is 'Rest of the World', then we can set $B_1 = 0$. This reflects the fact that German (or UK) exports are negligible compared with total output by the Rest of the World (i.e. the 'small country' assumption). Substitution of this assumption in (8.39)-(8.42) gives:

$$M = (I - A_1)^{-1} \tag{8.43}$$

$$Q = 0 \tag{8.44}$$

$$R = (I - A_2)^{-1} B_2 (I - A_1)^{-1} \tag{8.45}$$

$$S = (I - A_2)^{-1}. \tag{8.46}$$

Substituting from (8.43)-(8.46) into (8.37), the four components of CO_2 attributable to Region 1 become:

$$r^{1\prime} M y^{11} = r^{1\prime} (I - A_1)^{-1} y^{11} \tag{8.47}$$

$$r^{1\prime} Q y^{21} = 0 \tag{8.48}$$

$$r^{2\prime} R y^{11} = r^{2\prime} (I - A_2)^{-1} B_2 (I - A_1)^{-1} y^{11} \tag{8.49}$$

$$r^{2\prime} S y^{21} = r^{2\prime} (I - A_2)^{-1} y^{21}. \tag{8.50}$$

For Germany (or UK) (i.e. Region 1), we know A_1 and r^1. We do *not* know A_2 or r^2, for the Rest of the World. To make progress, we assume:

$$r^{1\prime} (I - A_1)^{-1} = r^{2\prime} (I - A_2)^{-1} \tag{8.51}$$

Substituting (8.51) into (8.49) and (8.50) gives:

Domestic CO_2 emission to meet Domestic final demand
$$= r^{1\prime} (I - A_1)^{-1} y^{11} \tag{8.52}$$

Foreign CO_2 emission to meet Domestic final demand

$$= r^{1'}(I - A_1)^{-1} B_2 (I - A_1)^{-1} y^{11} \qquad (8.53)$$

Foreign CO_2 emission to meet Imported final demand

$$= r^{1'}(I - A_1)^{-1} y^{21}. \qquad (8.54)$$

These three elements therefore sum to the total CO_2 emissions *attributable* to region 1; i.e. the CO_2 emissions for which region 1 is *responsible*) (Obviously this approach can be improved if estimates of A_2 and r^2 can be obtained.)

8.5 The Sensitivity of CO_2 Emissions to Changed Parameters

So far we have treated CO_2 emissions as dependent upon the structure of final demand (y), fuel use coefficients (C, P), the structure of inter-industry trading (A), etc. Further, we have assumed that these parameters are given, both in the sense of being known and being constant. For our analysis of the data, however, we must accept that the data we use are, like all data, subject to error, and therefore not perfectly known. Also, as we are interested in identifying policies that may reduce CO_2 emissions, we shall wish to consider how the effects of changing these parameters may be effective in achieving this aim.

An appreciation of the sensitivity of the level of total CO_2 emission to changes in the parameters and variables in the model is therefore important for three reasons.

Firstly, it provides a guide as to which components of the data need to be determined most accurately. This is because inaccuracy in a particular item of data matters much less if the elasticity of total CO_2 emissions, with respect to that item, is small than if it is large.

Second, elements in the technology matrices (A, C, P) which have high elasticities associated with them, identify potential for reducing CO_2 emissions through technical change.

Third, the sensitivities with respect to the individual components of final demand (y) provide an indication of the variations in the pattern of final demand which would be most worthwhile for reducing CO_2 emissions.

It is therefore important to find the sensitivity of CO_2 emissions to changes in the parameter values. At first sight this seems a formidable task, as there are approximately 2,600 parameters in the model used later in this study. Fortunately, as so often, some mathematical analysis makes the computation quite manageable.[4]

8.5.1 The Elasticities of CO_2 Emission with the Parameters

The usual measure of a sensitivity used by economists is, of course, the elasticity. This is the ratio of proportional changes for two variables. In order to calculate a particular elasticity, we need to know the partial derivative of one variable with respect to the other, and also the values of the variables at the point at which the elasticity is required. For example, the elasticity of total CO_2 emissions (C) with respect to one component of final demand (y_i) is given by:

$$\varepsilon^C_{y_i} \equiv \frac{\partial C/\partial y_i}{C/y_i}. \tag{8.55}$$

To calculate also the elasticities with respect to the technological coefficients (a_{ij}), production fuel use coefficients (c_{if}), and final demand fuel use (y_i), we use:

$$\varepsilon^C_{a_{ij}} \equiv \frac{\partial C/\partial a_{ij}}{C/a_{ij}}, \quad \varepsilon^C_{c_{if}} \equiv \frac{\partial C/\partial c_{if}}{C/c_{if}}, \quad \varepsilon^C_{p_{if}} \equiv \frac{\partial C/\partial p_{if}}{C/p_{if}}. \tag{8.56}$$

Here we use the usual definitions: f - fuel type; i, j - sector.

The first step in calculating these elasticities is to calculate the partial derivatives in the numerators.

---

[4] We are grateful to Simon Gay for suggesting this approach, rather than a direct computational strategy.

8.5.2 The Derivative of CO_2 Emission with Respect to a_{ij}

It will be recalled that total CO_2 emission is given by:

$$C = \mathbf{e'P'Zy} + (\mathbf{e'C'} + \mathbf{m'})(\mathbf{I} - \mathbf{A})^{-1}\mathbf{y}. \tag{8.57}$$

In order to calculate the elasticities with respect to the elements of the technology matrix, we need to find $\partial C / \partial a_{ij}$. We therefore define:

$$\mathbf{q'} \equiv \mathbf{e'C'} + \mathbf{m'} \tag{8.58}$$

$$\mathbf{L} \equiv \mathbf{I} - \mathbf{A}. \tag{8.59}$$

Then substituting (8.58) and (8.59) into (8.57), we can write:

$$C = \mathbf{e'P'Zy} + \mathbf{q'L}^{-1}\mathbf{y}. \tag{8.60}$$

Differentiating (8.60) with respect to a_{ij} gives:

$$\frac{\partial C}{\partial a_{ij}} = \mathbf{q'} \frac{\partial \mathbf{L}^{-1}}{\partial a_{ij}} \mathbf{y}. \tag{8.61}$$

To differentiate $\mathbf{L}^{-1}$ with respect to a_{ij} we write:

$$\mathbf{L}^{-1}\mathbf{L} = \mathbf{I} \tag{8.62}$$

Differentiating both sides of (8.62) with respect to a_{ij} gives:

$$\frac{\partial(\mathbf{L}^{-1}\mathbf{L})}{\partial a_{ij}} = \mathbf{0}. \tag{8.63}$$

By applying the product rule for differentiation, (8.63) can be written as:

8.5.2 The Derivative of CO_2 Emission with Respect to a_{ij}

$$\frac{\partial \mathbf{L}^{-1}}{\partial a_{ij}}\mathbf{L} + \mathbf{L}^{-1}\frac{\partial \mathbf{L}}{\partial a_{ij}} = 0. \tag{8.64}$$

Reorganising (8.64) gives:

$$\frac{\partial \mathbf{L}^{-1}}{\partial a_{ij}}\mathbf{L} = -\mathbf{L}^{-1}\frac{\partial \mathbf{L}}{\partial a_{ij}}. \tag{8.65}$$

Now if we post-multiply both sides of (8.65) by $\mathbf{L}^{-1}$ we get:

$$\frac{\partial \mathbf{L}^{-1}}{\partial a_{ij}} = -\mathbf{L}^{-1}\frac{\partial \mathbf{L}}{\partial a_{ij}}\mathbf{L}^{-1}. \tag{8.66}$$

Now $\mathbf{L} \equiv \mathbf{I} - \mathbf{A}$, so differentiation gives:

$$\frac{\partial \mathbf{L}}{\partial a_{ij}} = \mathbf{F}^{ij}. \tag{8.67}$$

Here we define the matrix $\mathbf{F}$ as:

$$F^{ij}_{rs} = \begin{cases} 1, & r=i, \ s=j \\ 0, & \text{otherwise} \end{cases}. \tag{8.68}$$

Substituting (8.67) into (8.66) we obtain:

$$\frac{\partial \mathbf{L}^{-1}}{\partial a_{ij}} = \mathbf{L}^{-1}\mathbf{F}^{ij}\mathbf{L}^{-1}. \tag{8.69}$$

Substituting (8.69) into (8.61), we obtain:

$$\frac{\partial C}{\partial a_{ij}} = \mathbf{q}'\mathbf{L}^{-1}\mathbf{F}^{ij}\mathbf{L}^{-1}\mathbf{y}$$
$$= \mathbf{q}'(\mathbf{I}-\mathbf{A})^{-1}\mathbf{F}^{ij}(\mathbf{I}-\mathbf{A})^{-1}\mathbf{y} \tag{8.70}$$
$$= \mathbf{q}'(\mathbf{I}-\mathbf{A})^{-1}\mathbf{F}^{ij}\mathbf{x}.$$

We now define:

$$\mathbf{v}' \equiv \mathbf{q}'(\mathbf{I} - \mathbf{A})^{-1}. \tag{8.71}$$

Substitution of (8.71) into (8.70) gives:

$$\frac{\partial C}{\partial a_{ij}} = \mathbf{v}'\mathbf{F}^{ij}\mathbf{x}. \tag{8.72}$$

Since the effect of the matrix $\mathbf{F}^{ij}$ is to pick out the required elements of the other matrices, (8.72) becomes:

$$\frac{\partial C}{\partial a_{ij}} = v_i x_j. \tag{8.73}$$

8.5.3 The Derivative of CO_2 Emissions with Respect to c_{if}

To estimate the elasticities with respect to the production fuel use coefficients, in the matrix $\mathbf{C}$, we use:

$$\frac{\partial C}{\partial c_{if}} = \mathbf{e}'\frac{\partial \mathbf{C}}{\partial c_{if}}(\mathbf{I} - \mathbf{A})^{-1}\mathbf{y}$$

$$= \mathbf{e}'\frac{\partial \mathbf{C}}{\partial c_{if}}\mathbf{x}. \tag{8.74}$$

By the same argument as in the above section, (8.74) becomes:

$$\frac{\partial C}{\partial c_{if}} = e_f x_i. \tag{8.75}$$

8.5.4 The Derivative of CO_2 Emissions with Respect to p_{if}

Final demand fuel use influence is accounted for by the matrix **P**, and the effect of varying the matrix is given by:

$$\frac{\partial C}{\partial p_{if}} = \mathbf{e}' \frac{\partial \mathbf{P}}{\partial p_{if}} \mathbf{Zy}. \tag{8.76}$$

We define:

$$\mathbf{w} \equiv \mathbf{Zy}. \tag{8.77}$$

Then from (8.76) and (8.77) the individual elements are given by:

$$\frac{\partial C}{\partial p_{if}} = e_f w_i. \tag{8.78}$$

8.5.5 The Derivative of CO_2 Emissions with Respect to y_i

he effect of variations in the composition of final demand (**y**) is analysed by noting that:

$$\frac{\partial C}{\partial y_i} = [\mathbf{e'P'Z} + (\mathbf{e'C'} + \mathbf{m'})(\mathbf{I} - \mathbf{A})^{-1}] \frac{\partial \mathbf{y}}{\partial y_i}. \tag{8.79}$$

We recall that:

$$\mathbf{v}' \equiv (\mathbf{e'C'} + \mathbf{m'})(\mathbf{I} - \mathbf{A})^{-1}. \tag{8.80}$$

We also define:

$$\mathbf{u}' \equiv \mathbf{e'P'Z}. \tag{8.81}$$

Substituting (8.80) and (8.81) into (8.79) we obtain:

$$\frac{\partial C}{\partial y_i} = [\mathbf{v}' + \mathbf{u}']\frac{\partial \mathbf{y}}{\partial y_i}. \tag{8.82}$$

So, as before, we can write (8.79) as:

$$\frac{\partial C}{\partial y_i} = v_i + u_i. \tag{8.83}$$

8.5.6 The Elasticities of CO_2 Emissions

The partial derivatives obtained above can now be used to calculate elasticities in the usual way. From (8.73), (8.75), (8.78) and (8.83) we have:

$$\varepsilon^C_{a_{ij}} = \frac{v_i x_j a_{ij}}{C} \tag{8.84}$$

$$\varepsilon^C_{c_{if}} = \frac{e_f x_i c_{if}}{C} \tag{8.85}$$

$$\varepsilon^C_{p_{if}} = \frac{e_i w_j p_{if}}{C} \tag{8.86}$$

$$\varepsilon^C_{y_i} = \frac{(v_i + u_i) y_i}{C}. \tag{8.87}$$

These elasticities can all be calculated relatively easily.

8.5.7 The Elasticities of CO_2 Emissions with Respect to $a_{.j}$ and $a_{i.}$

As the input-output data we shall use in later chapters uses a 47-sector classification, we shall be able to derive $47 \times 47 = 2209$ $\varepsilon^C_{a_{ij}}$ elasticities.

8.5.7 The Elasticities of CO_2 Emissions with Respect to $a_{.j}$ and $a_{i.}$

This is rather too rich a data set to be able to deal with adequately. Also, it would be useful for policy purposes to look more directly at technology elasticities related to one sector at a time.

We therefore introduce two new technology elasticities. The first corresponds to the technology of *inputs* to a sector; the second to the technology of *outputs* from a sector. We summarise these input and output relationships simply with the sums of the technology coefficients. It should be noted that this summation is only permissible if the coefficients are defined upon a *value* input-output table. If they are defined upon a *quantity* table, then the coefficients are all in different physical units, so they cannot be added.

We define the coefficient sums as follows:

$$a_{.j} \equiv \sum_i a_{ij}, \qquad a_{i.} \equiv \sum_j a_{ij}. \tag{8.88}$$

We can now define the sectoral technology elasticities as follows:

$$\varepsilon^C_{a_{.j}} \equiv \frac{\partial C/\partial a_{.j}}{C/a_{.j}}, \qquad \varepsilon^C_{a_{i.}} \equiv \frac{\partial C/\partial a_{i.}}{C/a_{i.}}. \tag{8.89}$$

Fortunately, it is straightforward to express these sectoral technology elasticities as weighted sums of the single coefficient elasticities, $\varepsilon^C_{a_{ij}}$. We note:

$$\frac{\partial C}{\partial a_{.j}} = \frac{1}{\frac{\partial a_{.j}}{\partial C}} = \frac{1}{\frac{\partial \sum_i a_{ij}}{\partial C}} = \frac{1}{\sum_i \frac{\partial a_{ij}}{\partial C}}. \tag{8.90}$$

Now from (8.56) we have:

$$\frac{\partial a_{ij}}{\partial C} = \frac{1}{\varepsilon^C_{a_{ij}}} \frac{a_{ij}}{C}. \tag{8.91}$$

Substituting from (8.91) in (8.90) gives:

$$\frac{\partial C}{\partial a_{.j}} = \frac{1}{\sum_i \frac{1}{\varepsilon^C_{a_{ij}}} \frac{a_{ij}}{C}}. \tag{8.92}$$

Substitution of (8.92) in (8.89) gives:

$$\varepsilon^C_{a_{\cdot j}} = \frac{a_{\cdot j}/C}{\sum_i \frac{1}{\varepsilon^C_{a_{ij}}} \frac{a_{ij}}{C}}. \tag{8.93}$$

Equation (8.93) reorganises to give:

$$\varepsilon^C_{a_{\cdot j}} = \frac{1}{\sum_i \frac{a_{ij}}{a_{\cdot j}} \frac{1}{\varepsilon^C_{a_{ij}}}}. \tag{8.94}$$

By similar reasoning we obtain:

$$\varepsilon^C_{a_{i \cdot}} = \frac{1}{\sum_j \frac{a_{ij}}{a_{i \cdot}} \frac{1}{\varepsilon^C_{a_{ij}}}}. \tag{8.95}$$

8.6 Conclusions

In this chapter we have undertaken a considerable amount of analysis. We have shown that the total CO_2 emissions from an economy can be represented in input-output form. We have seen that this allows us to deal with the direct and indirect CO_2 attribution of CO_2 emissions between producing sectors.

We have also seen how imports and exports can be included in the input-output framework, to allow us to distinguish between 'CO_2 emission' and 'CO_2 responsibility'.

Finally, we have shown that CO_2 emissions can be related to various parameters, using an elasticity approach. In the rest of this study we apply these results to build upon our understanding of how economic structural change affects CO_2 emissions, as discussed in Chapters 4 and 5.

Part IV

Data Analysis

9 German and UK Input-Output Data for Studying CO_2 Emissions

9.1 Introduction

In Chapter 8 we established the theory of input-output analysis, as applied to CO_2 emissions. We now begin to apply this theory, by establishing the data base we use, and making a preliminary analysis of the data.

In Section 9.2 we outline the data requirements for this input-output study. Section 9.3 concerns the data collection and processing undertaken. The data sources used for establishing the input-output database are detailed in Section 9.4. As an example of the data collected, the data for Germany in 1988 are discussed and analysed in Section 9.5. Finally, Section 9.6 contains some conclusions.

In the next chapter we shall calculate the other measures derived in Chapter 8; i.e. the difference decomposition of CO_2 emissions, and the distinction between CO_2 emissions and CO_2 'responsibility' by countries.

9.2 The Data Requirements

In Chapter 8 we derived our fundamental input-output equation for total CO_2 emission by an n-sector economy. We recall that this is:

$$C = [\mathbf{e'P'Z} + (\mathbf{e'C'} + \mathbf{m'})(\mathbf{I} - \mathbf{A})^{-1}]\mathbf{u}Y. \tag{8.16}$$

Here we describe the total emission of CO_2, C, for an n-sector economy, in terms of the following:

- **e** - The 3-vector of CO_2 emission per unit of fossil-fuel use (for the three fuels: gas, coal and oil).

- **P** - The $(n \times 3)$ matrix of household fuel requirements per unit of final demand, for each producing sector, and for each fossil-fuel. (This matrix has only three non-zero elements.)

- **Z** - The $(n \times n)$ matrix of weights, to exclude from consideration the export and stock building of fuels. (This matrix has only three non-zero elements.)

C - The ($n \times 3$) matrix of production fuel requirements per unit of total demand, for each producing sector, and for each fossil-fuel.

m - The n-vector of CO_2 emissions per unit of total demand, attributable to non-fossil fuel sources, for each producing sector.

A - The ($n \times n$) matrix of technological coefficients, relating the inputs by each producing sector to the total outputs by the sectors.

B - The ($n \times n$) matrix of coefficients, relating imported inputs to each producing sector to the total outputs by the sectors.

u - The n-vector of final demand shares in GDP for each producing sector.

Y - The GDP of the economy

Finally, it should be noted that, throughout this study, the term 'Germany' is used as an abbreviation for 'Federal Republic of Germany'. It should also be stressed that during the period of this study the Federal Republic expanded to include the former German Democratic Republic, with consequent extensive economic alterations. In this study, for reasons of data availability, by 'Germany' we mean the Federal Republic prior to its enlargement in 1990.

9.3 Data Collection and Processing

In order to obtain an estimate of total CO_2 emissions, C, in its disaggregated form, it is necessary to estimate e, P, Z, C, m, A, B and y from published statistics. The data necessary was generally not directly available in the appropriate, or consistent, form, so a considerable amount of data processing was required before the analysis could commence. The processing necessary was in two parts.

1. The input-output tables used have been aggregated to a consistent 47 sector basis. This was found to be the minimum aggregation of the German data (58 sectors) and the UK data (90, 100 or 102 sectors), which gave consistent sectoral definitions.

2. The necessary energy statistics on fossil fuel use in the two countries had to be checked for their internal consistency, and adjusted where necessary.

We first turn to a discussion of the aggregation procedure used.

9.3.1 The Aggregation of the Input-Output Tables

The input-output tables used in this study are as shown below.

Germany   1978 1980 1982 1984 1986 1988
UK   1968 1974 1979 1984

While the German input-output data is collected frequently (every two years) and published quite rapidly (less than two years), this is not the case in the UK. There the data is collected only every five or six years, and takes at least two years to be published. A consequence of this disparity in data collection and publication is that for the UK we have a relatively sparse set of tables (four tables), but over a long period (25 years). On the other hand, for Germany we have six tables, but spanning only 15 years. Further, while the latest German input-output tables are for 1988, the latest for the UK are 1984. (At the time of writing, the 1989 UK input-output tables were under preparation, but were unavailable for this study.)

To make meaningful comparisons between the input-output tables, over time and between countries, the tables had to be aggregated to a consistent sectoral basis. Two major difficulties were encountered in the aggregation procedure.

First, the definitions of some sectors seemed, at first sight, to be identical for Germany and the UK. However, close checking of the exact sectoral definitions, in terms of the Standard Industrial Classification (SIC) elements involved, often showed there were differences. Very close study of the exact SIC elements involved in each sector was necessary before an aggregation scheme could be devised. Finally a scheme of 45 sectors was devised, which was the least possible aggregation, consistent with the data sources.

The second difficulty concerned the Electricity Supply sector. In both Germany and the UK this is comprised of three quite separate elements:

1. Electricity generation using fossil fuels.

2. Electricity generation using non-fossil fuels (nuclear and hydro).

3. Electricity distribution.

For the purposes of determining the scope for the reduction of CO_2 emission, it is obviously important that these three elements be separated. The disaggregation of this sector was achieved as follows.

The two generating sectors were assumed to sell all of their output to the distribution sector, which has no other intermediate purchases. The fuel inputs to the Electricity sector were attributed entirely to Fossil-Fuel Generation, and all other inputs were split between the two generating sectors, in proportion to their total output. All purchases of electricity by the remaining sectors, and by final demand, were attributed to the Electricity Distribution sector.

The disaggregation of the original Electricity sector into three sectors meant that the final input-output structure had 47 sectors.

9.4 Data Sources Used

We now describe the data sources used to construct the elements discussed in Section 9.3. We also outline the data processing necessary. This is done at some length, but as much data manipulation was necessary, it is appropriate that the reader be fully aware of our assumptions and estimations.

9.4.1 Production of the A Matrices for Germany

The source of the German A matrices used was the flows matrices (commodity analysis of purchases by sector from domestic production) from the *Statistisches Bundesamt* [1989]. These were aggregated appropriately.

The flows matrices thus obtained were adjusted for price variations over time, using sectoral price indices, obtained from the *Statistisches Bundesamt* [1989]. The basis year was 1980.

The 47 sector A matrix was then derived in the normal way, by dividing inter-industry flows by total outputs.

9.4.2 Production of the A Matrices for the UK

The methodology followed was the same as for the German data. The basis for the production of the 47 sector A matrix used was the flow matrices from the 1968, 1974, 1979 and 1984 *Input-Output Tables* [Central Statistical Office 1973, 1980, 1983, 1988]. The 47 sector A matrices were derived by dividing inter-industry flows by total outputs.

All monetary quantities were deflated by price indices corresponding to the individual sectors. The basis year was 1980, as for Germany. The principal sources for these price indices were the *Annual Abstract of Statistics* [Central Statistical Office 1980, 1984, 1987]. Also, the *Digest of United Kingdom Energy Statistics* [Department of Energy 1989] was used to determine some of these price indices.

9.4.3 Production of the C Matrices for Germany

The C matrix contains data on primary fuel use by sector, measured in physical units and disaggregated into three types of fuel: solid, liquid and gas. This data on energy use in Germany, by fuel type, is easily available. It is recorded, by input-output sector, for intermediate demand, and also for final demand. This data was taken from the *Statistisches Bundesamt* [1989].

9.4.4 Production of the C Matrices for the UK

In the UK, the data on fuel use is, unfortunately, not collected together in one place. As a result, a very lengthy process of data collation was necessary.

The starting points were the *1974 Report of the Censuses of Production* [Department of Industry 1975], *1979 Census of Production and Purchases Inquiry* [Department of Industry 1980] and the *1984 Purchases Inquiry* [Department of Trade and Industry 1987]. These covered, with some exceptions, all of the activity headings within divisions 1 to 4 of the Standard Industrial Classifications (Revised 1980). The same classification was assumed for the data for the years 1974 and 1979. For a great number of the activities, data is published on fuel use, measured both in value terms and in most cases physical units, divided into three types of solid fuel, three (or two) types of liquid fuel and two types of gas. To convert the different physical units to one basis unit, the conversion factors, published in *Energy Statistics* [Department of Energy 1989], were used.

For activities where no physical units were recorded, the *Input-Output Tables* [Central Statistical Office 1973, 1980, 1983, 1988] and *Energy Statistics* [Department of Energy 1981, 1989] were used. In *Energy Statistics* [Department of Energy 1981, 1989] are published the energy consumption by final users, and a classification of consumers on the basis of the Standard Industrial Classification.

The data on primary fuel use by sector, which result from the *Purchases Inquiry* [Department of Trade and Industry 1987], were adjusted with the figures from *Energy Statistics* [Department of Energy 1981, 1989]. Some sectors were aggregated in the *Energy Statistics* data. To achieve disaggregation of energy use by those sectors, data from the input-output tables were used. The aggregated energy use was disaggregated proportionally to the value purchased by each sector of the corresponding energy type.

Similar action was taken for the Transport sector, because in this sector it must be taken into consideration that the figure for energy use contains transport services (intermediate demand) and transport by consumers (final demand).

To obtain aggregate fuel use consistent with the Inland Consumption of primary fuel, published in *Energy Statistics* [Department of Energy 1981, 1989], a scaling factor was used. This scaling factor was the ratio of the calculated fuel use (results from above) and the published fuel use (*Energy Statistics*). Multiplication by this ratio of the calculated fuel use for each sector gave adjusted fuel use, which overall sums to the known national fuel use.

Division of each element of the (3×47) matrix thus obtained, by the corresponding element of the total output vector, produced the (3×47) matrix **C**, which shows primary fuel use per unit of total output.

9.4.5 Production of the P Matrices for Germany

The model also requires a matrix, **P**, which contains the household fuel use per unit final demand. The three non-zero elements of this matrix were estimated using data published in the *Statistisches Bundesamt* [1989]. In the same way as the elements of the **C** matrix depend on the technology of production, the elements of **P** depend on the composition of final demand. For example, the liquid fuel element is an aggregation of fuel oil, used for heating, and petroleum products, used as motor fuel. The three non-zero elements of this matrix were then divided by the corresponding element of the final demand, to give the **P** matrix. In this case the final demand does not contain the stocks change and the exports. These figures were omitted as interest is directed towards only that fuel which were burnt.

9.4.6 Production of the P Matrices for the UK

For the UK, the three non-zero elements of this matrix were estimated using data published in *Energy Statistics* [Department of Energy 1981, 1989]. These figures were scaled, as for the treatment of the UK C matrix. The derivation of the P matrix was as for Germany.

9.4.7 Production of the Z Matrices for Germany

The Z matrix is a (47×47) diagonal matrix, with only three non-zero elements, for gas, coal and oil. The elements are the ratios between the final demand for fuels consumed domestically, and the total final demand, including exports and stock building. These were derived from the input-output tables.

9.4.8 Production of the Z Matrices for the UK

For the UK Z matrix, the same methodology as for Germany was used.

9.4.9 Production of the e Vector for Germany

The e vector is a vector of CO_2 emission per unit of fuel burnt. The assumptions used here were as in Section 3.2; i.e.:

1. All gas consists of methane, and generates 5.4 tonnes CO_2 per Ktherm.

2. Solid fuel is 60% carbon by weight, and generates 2.2 tonnes CO_2 per tonne of fuel.

3. All liquid fuel consists of molecules with eight carbon atoms, and generates 3.13 tonnes CO_2 per tonne of fuel.

9.4.10 Production of the e Vector for the UK

The e vector used for the UK was the same as for Germany.

9.4.11 Production of the m Vectors for Germany

Representative firms in brick making (with carbonaceous clay), glass making and waste disposal were approached, and their data used to estimate m.

9.4.12 Production of the m Vectors for the UK

For the UK m vectors, the methodology as for Germany was used.

9.4.13 Production of the B Matrices for Germany

The B matrices for Germany were derived from the published input-output tables, using the same technique as for establishing the A matrices.

9.4.14 Production of the B Matrices for the UK

For the UK B matrices, the methodology as for Germany was used.

9.4.15 Production of the u Vectors and Y for Germany

The elements of u vector are simply the ratios of the sectoral final demands to GDP (Y). Now the definition of GDP is that it is total final demand, summed over all sectors. Hence, Y is found by summing the final demand elements, and u is then derived by division of the elements of y by Y.

9.4.16 Production of the u Vectors and Y for the UK

The same methodology was used as for Germany.

9.4.17 Further Adjustment of the UK Data

One further adjustment was carried out to the UK final demand of the coal sector for 1984. This adjustment was necessitated by the coal strike of 1984-5. In that year coal stocks were decreasing very strongly and final demand for the Coal sector was therefore negative. The adjusted figure was calculated by consideration of the time-trend of Coal final demand. The adjustment was accomplished by a reduction of the stock change in monetary units. It follows from this that the total output of the coal sector for 1984 also had to be changed.

9.5 The Structure of CO_2 Emission: Germany 1988

We begin by a straightforward input-output analysis of the data, following the methodology discussed in Chapter 8. We present illustrative data for Germany in 1988. As the full data set, for both countries and all years, is very extensive, this is not presented.[1]

9.5.1 The Nature of Input-Output Data

Before moving on to an analysis of the data, a word of caution is in order. The amount of data generated by input-output analysis is enormous. For example, simply to store the data we have generated takes over 30 megabytes of computer memory. In such work, any particular statistic can be generated for each sector, both countries and every year. For this study, this gives 940 cases. Further, it is possible to relate each sector to every other sector for many measures, giving 44,180 cases. If all possible inter-country and inter-year comparisons are also sought this takes the number of potential cases up to 1.95×10^9. The human brain (at least ours!) cannot cope with this amount of information.

---

[1] This data set is available to interested researchers, on application to the first author.

158   9 German and UK Input-Output Data for Studying CO_2 Emissions

It is imperative that the information generated be somehow condensed so that it can be comprehended, and thus allow policy conclusions to be drawn. We therefore purposely present a relatively condensed set of statistics, and ask the reader to accept that these are appropriate for the analysis undertaken. However, we do not wish to suggest that we have exploited the full potential of our data set; many further areas of work remain.

9.5.2 The Basic Data Used in the Analysis

We begin with an outline of the basic data used as an input to our calculations. In Table 9.1, the Total Output and Final Demand columns are simply the equivalent sections of the German 1988 Input-Output Tables, aggregated to the 47-sector level as described above. It will be noticed that the two electricity generating sectors have zero final demand.

The next column contains the three non-zero elements of the **P** matrix, for the three fuels. These three elements represent the fuel delivered to final consumers, per million DM of final demand. (The units for these components differ between fuels; they are as for the **C** matrix, to be described next.)

The final three columns record primary fuel use per unit of gross output by sector (i.e. matrix **C**). These represent the fuel delivered to each sector, per million DM of total output. Once again, the entries for the electricity sectors are worth commenting on. The distribution side of electricity is recorded as using no fossil fuel at all, which is clearly an underestimate. The same is true of Other Electricity Generation.

9.5.3 The CO_2 Intensities

We now move on to a consideration of the CO_2 intensities we can derive from the basic data. Table 9.2 shows the various contributions to total CO_2 emission.

9.5.3 The CO$_2$ Intensities

Sector	Final Demand MDM	Total Output MDM	P Units	C(gas) Kth/ MDM	C(coal) Mt/ MDM	C(oil) Mt/ MDM
1 Agriculture	15990.6	64148.7	0.00	0.40	0.62	63.06
2 Forestry & fishing	2552.3	8301.6	0.00	6.33	1.93	77.33
3 Electricity; fossil generation	0.0	36552.9	0.00	89.10	1975.56	93.29
4 Electricity; other generation	0.0	24095.7	0.00	0.00	0.00	0.00
5 Electricity; distribution	22129.8	60648.5	0.00	0.00	0.00	0.00
6 Gas	9519.7	18531.6	777.80	0.00	0.00	1.51
7 Water	468.0	5502.5	0.00	0.00	0.00	5.09
8 Coal extraction, coke ovens, etc	1378.6	20659.9	10.56	3.06	1401.69	2.08
9 Extraction of metalliferous ores	1274.9	4919.1	0.00	19.50	4.05	6.30
10 Extraction of mineral oil and gas	374.4	4202.0	0.00	1.18	0.00	1.67
11 Chemical products	86824.4	160105.2	0.00	18.40	15.81	26.55
12 Mineral oil processing	30774.1	60338.2	1.78	2.18	0.00	19.51
13 Processing of plastics	16212.7	43422.0	0.00	2.10	0.07	3.87
14 Rubber products	5778.5	11518.2	0.00	7.49	4.08	7.47
15 Stone, clay, cement	5719.8	31031.6	0.00	16.22	62.16	33.68
16 Glass, ceramic goods	6323.0	14669.4	0.00	36.34	0.76	35.72
17 Iron and steel, steel products	29191.9	116232.9	0.00	11.71	10.67	9.39
18 Non-ferrous metals	10134.4	25532.3	0.00	9.01	3.82	5.13
19 Foundries	2079.1	13829.3	0.00	10.81	0.69	10.56
20 Production of steel etc	14758.1	22546.1	0.00	1.39	0.27	7.05
21 Mechanical engineering	94881.3	122948.2	0.00	1.55	0.26	7.03
22 Office machines	17710.4	18888.3	0.00	1.41	0.00	1.91
23 Motor vehicles	128463.6	160025.1	0.00	2.09	0.58	2.67
24 Shipbuilding	3608.9	4320.4	0.00	2.14	0.00	6.25
25 Aerospace equipment	9436.4	10881.8	0.00	2.35	0.37	1.93
26 Electrical engineering	92348.4	134171.6	0.00	0.70	0.18	5.10
27 Instrument engineering	15132.3	20000.8	0.00	0.76	0.00	3.65
28 Engineers' small tools	24078.7	41313.3	0.00	2.87	0.15	8.40
29 Music instruments, toys, etc	6999.2	7474.9	0.00	0.76	0.00	4.01
30 Timber processing	2243.4	10924.5	0.00	1.72	5.62	24.71
31 Wooden furniture	22057.5	28373.9	0.00	0.25	0.29	12.48
32 Pulp, paper, board	6736.2	17324.9	0.00	19.54	25.49	34.23
33 Paper and board products	7126.3	19137.0	0.00	3.49	0.00	6.27
34 Printing and publishing	3951.9	28075.3	0.00	1.67	0.00	3.56
35 Leather, leather goods, footwear	6001.6	6906.3	0.00	1.13	0.26	8.25
36 Textile goods	18984.0	31480.6	0.00	8.15	2.73	11.79
37 Clothes	18469.0	20148.8	0.00	0.37	0.15	6.30
38 Food	102358.2	142453.0	0.00	5.16	1.30	12.99
39 Drink	15051.5	22980.4	0.00	7.88	0.96	19.02
40 Tobacco	13171.7	14014.7	0.00	0.51	0.29	1.64
41 Construction	170319.5	203825.3	0.00	0.07	0.24	11.94
42 Trade wholesale & retail	188509.2	273288.2	0.00	1.39	0.08	15.35
43 Traffic & transport services	51118.5	112030.0	0.00	0.11	0.00	93.24
44 Telecommunications	27339.9	51874.6	0.00	0.30	0.08	5.47
45 Banking, finance, insurance, etc	173839.6	285890.3	0.00	0.23	0.00	1.52
46 Hotels, catering, etc	37721.6	55698.0	0.00	3.02	0.22	12.21
47 Other services	219485.3	499037.6	0.00	0.62	0.01	7.14

Table 9.1. Value of output and fuel use: Germany 1988.

Sector	e'P'Z Ktonnes CO_2/MDM	e'C' Ktonnes CO_2/MDM	e'C'(A+.) Ktonnes CO_2/MDM	Total Ktonnes CO_2/MDM
1 Agriculture	0.00	0.20	0.19	0.39
2 Forestry & fishing	0.00	0.28	0.17	0.45
3 Electricity; fossil generation	0.00	5.12	1.46	6.58
4 Electricity; other generation	0.00	0.00	0.38	0.38
5 Electricity; distribution	0.00	0.00	4.12	4.12
6 Gas	4.17	0.00	0.08	4.26
7 Water	0.00	0.02	0.74	0.75
8 Coal extraction, coke ovens, etc	5.45	3.11	1.22	9.78
9 Extraction of metalliferous ores	0.00	0.13	0.26	0.39
10 Extraction of mineral oil and gas	0.00	0.01	0.20	0.21
11 Chemical products	0.00	0.22	0.27	0.49
12 Mineral oil processing	4.92	0.07	0.06	5.05
13 Processing of plastics	0.00	0.02	0.21	0.23
14 Rubber products	0.00	0.07	0.19	0.26
15 Stone, clay, cement	0.00	0.33	0.35	0.68
16 Glass, ceramic goods	0.00	0.31	0.25	0.56
17 Iron and steel, steel products	0.00	0.12	0.62	0.73
18 Non-ferrous metals	0.00	0.07	0.37	0.44
19 Foundries	0.00	0.09	0.30	0.40
20 Production of steel etc	0.00	0.03	0.18	0.21
21 Mechanical engineering	0.00	0.03	0.16	0.19
22 Office machines	0.00	0.01	0.12	0.13
23 Motor vehicles	0.00	0.02	0.19	0.21
24 Shipbuilding	0.00	0.03	0.20	0.23
25 Aerospace equipment	0.00	0.02	0.09	0.11
26 Electrical engineering	0.00	0.02	0.12	0.14
27 Instrument engineering	0.00	0.02	0.11	0.12
28 Engineers' small tools	0.00	0.04	0.21	0.25
29 Music instruments, toys, etc	0.00	0.02	0.13	0.15
30 Timber processing	0.00	0.10	0.25	0.35
31 Wooden furniture	0.00	0.04	0.18	0.23
32 Pulp, paper, board	0.00	0.27	0.46	0.73
33 Paper and board products	0.00	0.04	0.24	0.28
34 Printing and publishing	0.00	0.02	0.19	0.21
35 Leather, leather goods, footwear	0.00	0.03	0.11	0.14
36 Textile goods	0.00	0.09	0.20	0.29
37 Clothes	0.00	0.02	0.13	0.15
38 Food	0.00	0.07	0.25	0.32
39 Drink	0.00	0.10	0.18	0.29
40 Tobacco	0.00	0.01	0.04	0.05
41 Construction	0.00	0.04	0.14	0.18
42 Trade wholesale & retail	0.00	0.06	0.10	0.16
43 Traffic & transport services	0.00	0.29	0.15	0.44
44 Telecommunications	0.00	0.02	0.04	0.06
45 Banking, finance, insurance, etc	0.00	0.01	0.07	0.08
46 Hotels, catering, etc	0.00	0.05	0.21	0.27
47 Other services	0.00	0.03	0.06	0.08

Table 9.2. CO_2 intensities: Germany 1988.

The first column corresponds to $e'P'Z$, the direct consumption demand for fossil fuels. This column shows the tonnes of CO_2 emitted per million DM of demand by consumers for fuel. This column contains only three non-zero elements, one for each type of fossil fuel. We see that the emission of CO_2 through the purchase of 1 DM of household fuel, is almost the same for the three fuels (gas: 4.17; coal: 5.45; oil: 4.92).

The second column corresponds to $e'C$, the direct production demand for fuel. This column shows the tonnes of CO_2 emitted directly by each sector, per million DM of final demand for the output of that sector. As would be expected, Fossil Fuel Electricity Generation (Sector 3) has the highest direct CO_2 emission per DM. The high value for coal extraction (Sector 8) is also noteworthy, while the low value for Iron and Steel (Sector 17) is also interesting. Otherwise the levels and ranking of these figures seems reasonable[2].

The third column corresponds to $e'C(A + A^2 + ...)$, the indirect production demand for fuel. This is the tonnes of CO_2 emitted throughout the rest of the economy for each sector, per million DM of final demand for the output of that sector. It is noteworthy that the great majority of sectors are 'responsible' for much more CO_2 production indirectly than directly. There are only seven sectors which do not follow this pattern. These are sectors 1, 2, 3, 8, 12, 16 and 43. This preponderance of indirect over direct CO_2 emission by sector holds for both countries, in all the years considered. This indicates how crucial it is to use an approach which takes economic interrelations into account when analysing CO_2 production.

The fourth column is the sum of the first three, and is the total 'CO_2 intensity' per million DM final demand, for each sector.

9.5.4 Attributed CO_2 Emissions

We now consider the attribution of CO_2 emissions, using our input-output approach. This is presented in Table 9.3, which shows the CO_2 emission

---

[2] On comparing the Iron and Steel and Coal Extraction direct intensities for all the German and UK tables, it was seen that for the UK the Coal Extraction figures are much lower than for Germany, while the Iron and Steel figures are much higher. This discrepancy is almost certainly because in the German data the burning of coke in steel making is associated with the coke production, which is in the Coal sector. In the UK, the burning of the coke is associated with the Iron and Steel sector.

that takes place, for each sector, derived from the CO_2 intensities in Table 9.2. In each case the corresponding CO_2 intensity has been multiplied by the final demand for goods and services, for each sector.

Below these four columns are the total CO_2 emission attributable to direct consumption demand, direct production demand, and indirect production demand for fuels, both in absolute terms and as percentages of the overall total. We see once more the importance of the indirect demand for fuels in the production of CO_2. Nearly 54% of German CO_2 emission in 1988 is attributable to the indirect demand for fossil fuels. Slightly more than 30% of the emission is directly attributable to household demand for fossil fuels. These figures are remarkably stable, over time and between the two countries, as is shown in Figure 9.1, for both countries and all years.

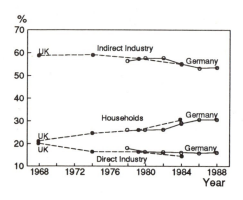

Figure 9.1. The shares of CO_2 emissions for Germany and the UK (1968-88).

Figure 9.1 shows the similarity of level and trend of the shares of CO_2 emission, for Germany and the UK. Where the UK and German data are for different years, they interpolate each other very well. For the one year they share (1984), the coincidence is very close.

It is also apparent that there are trends in the attributed CO_2 emissions. The trend of the share of households, for both Germany and the UK, is clearly upwards. Conversely, the trend of the share of indirect industrial CO_2 emissions is clearly downwards, for both countries. There is no clear trend for the share of direct industrial CO_2 emissions. (These finding are in line with the results found, by other methods, in Section 5.5.)

That the household trend should be upwards is perhaps unsurprising, reflecting the greater use of personal transport by households. The downward trend for indirect industrial CO_2 emissions, however, *is* surprising. The usual view regarding economic interconnectedness, through inter-industry trading, is that it is increasing, which would suggest an increasing trend for this category of CO_2 emissions. Two possible explanations suggest themselves here. First, the usual assumption of the increasing 'connectedness' of modern economies may be incorrect.

9.5.4 Attributed CO_2 Emissions

	Sector	e'P'Zy Ktonnes CO_2	e'C'y Ktonnes CO_2	e'C'(A+.)y Ktonnes CO_2	Total Ktonnes CO_2
1	Agriculture	0.0	3212.4	3003.2	6215.6
2	Forestry & fishing	0.0	715.9	440.4	1156.3
3	Electricity; fossil generation	0.0	0.0	0.0	0.0
4	Electricity; other generation	0.0	0.0	0.0	0.0
5	Electricity; distribution	0.0	0.0	91169.3	91169.3
6	Gas	39683.4	45.0	808.2	40536.7
7	Water	0.0	7.5	345.8	353.3
8	Coal extraction, coke ovens, etc	7516.1	4282.9	1682.4	13481.4
9	Extraction of metalliferous ores	0.0	170.7	332.8	503.5
10	Extraction of mineral oil and gas	0.0	4.3	74.5	78.8
11	Chemical products	0.0	18859.9	23523.6	42383.5
12	Mineral oil processing	151479.5	2241.9	1702.5	155423.9
13	Processing of plastics	0.0	382.8	3412.5	3795.4
14	Rubber products	0.0	420.7	1094.6	1515.3
15	Stone, clay, cement	0.0	1886.0	1979.1	3865.2
16	Glass, ceramic goods	0.0	1958.5	1564.6	3523.0
17	Iron and steel, steel products	0.0	3389.0	18042.2	21431.3
18	Non-ferrous metals	0.0	741.3	3751.6	4492.9
19	Foundries	0.0	193.3	632.7	826.0
20	Production of steel etc	0.0	444.9	2673.8	3118.7
21	Mechanical engineering	0.0	2932.3	15125.8	18058.2
22	Office machines	0.0	240.5	2127.3	2367.9
23	Motor vehicles	0.0	2688.8	24594.0	27282.8
24	Shipbuilding	0.0	112.2	718.6	830.9
25	Aerospace equipment	0.0	184.4	890.1	1074.5
26	Electrical engineering	0.0	1856.9	10631.2	12488.1
27	Instrument engineering	0.0	235.3	1602.3	1837.5
28	Engineers' small tools	0.0	1014.1	4943.5	5957.6
29	Music instruments, toys, etc	0.0	116.7	922.3	1039.0
30	Timber processing	0.0	222.2	551.8	774.0
31	Wooden furniture	0.0	905.4	4060.4	4965.8
32	Pulp, paper, board	0.0	1810.1	3088.0	4898.1
33	Paper and board products	0.0	274.1	1734.0	2008.1
34	Printing and publishing	0.0	79.7	731.6	811.3
35	Leather, leather goods, footwear	0.0	195.1	658.5	853.6
36	Textile goods	0.0	1649.2	3872.4	5521.6
37	Clothes	0.0	407.3	2323.6	2731.0
38	Food	0.0	7305.5	25272.5	32578.1
39	Drink	0.0	1568.3	2726.1	4294.4
40	Tobacco	0.0	112.0	510.6	622.6
41	Construction	0.0	6520.6	23752.2	30272.8
42	Trade wholesale & retail	0.0	10506.4	18879.3	29385.7
43	Traffic & transport services	0.0	14949.6	7637.6	22587.2
44	Telecommunications	0.0	517.6	1127.7	1645.3
45	Banking, finance, insurance, etc	0.0	1041.8	12669.1	13710.9
46	Hotels, catering, etc	0.0	2073.8	7943.9	10017.7
47	Other services	0.0	5647.8	12408.7	18056.5
	Total	198679.0	104124.7	347737.3	650540.9
	%	30.5	16.0	53.5	100

Table 9.3. CO_2 emissions: Germany 1988.

Second, connectedness may be 'shifting' from *intra*-country relationships to *inter*-country connectedness. Such inter-country connectedness would not be picked-up by the analysis so far. We return to this issue in the next chapter.

9.5.5 CO_2 Intensities and Emissions by Sector

To get a clearer view of the 'guilty' sectors in CO_2 emission, the total CO_2 intensity is shown in Table 9.4, ranked by size. The upper ranking is much as one might expect, with energy production and distribution at the top, except for Other (non-fossil fuel) Electricity Generation (Sector 4). In the middle ranks we find Drink (Sector 39) above Mechanical Engineering (Sector 21), which is rather surprising. Even more surprising is to find that Agriculture (Sector 1) is more CO_2 intensive than Shipbuilding (Sector 24) and Electrical Engineering (Sector 26).

Finally, in Table 9.5 we also rank total CO_2 emission, by sector. We remind the reader that this relates to fuels use by consumers directly, by sectors directly, and by sectors indirectly. Study of the complete data set, for both countries, shows that there is a reasonable consistency between countries and over time, of the sectoral rankings of both CO_2 intensity and total CO_2 emission.

9.5.6 CO_2 Emission Elasticities

In Section 8.5 we developed the principles of CO_2 elasticities. In particular, we there derived the following elasticities.

1. Final demand elasticities of CO_2 emission. These relate CO_2 emission to the final demand vector (y).

2. Fuel use elasticities of CO_2 emission. These relate CO_2 emissions to the matrices of fuel use coefficients (P and C).

3. Intermediate trading elasticities of CO_2 emission. These relate CO_2 emissions to the matrix of intermediate demand coefficients (A).

We also established the elasticities of CO_2 emissions with regard to the row sums and column sums of the A matrix, as summary measures.

9.5.6 CO_2 Emission Elasticities

Rank		Sector	Intensity tonnes CO_2/ KDM
1	8	Coal extraction, coke ovens, etc	9.78
2	3	Electricity; fossil generation	6.58
3	12	Mineral oil processing	5.05
4	6	Gas	4.26
5	5	Electricity; distribution	4.12
6	7	Water	0.75
7	17	Iron and steel, steel products	0.73
8	32	Pulp, paper, board	0.73
9	15	Stone, clay, cement	0.68
10	16	Glass, ceramic goods	0.56
11	11	Chemical products	0.49
12	2	Forestry & fishing	0.45
13	18	Non-ferrous metals	0.44
14	43	Traffic & transport services	0.44
15	19	Foundries	0.40
16	9	Extraction of metalliferous ores	0.39
17	1	Agriculture	0.39
18	4	Electricity; other generation	0.38
19	30	Timber processing	0.35
20	38	Food	0.32
21	36	Textile goods	0.29
22	39	Drink	0.29
23	33	Paper and board products	0.28
24	46	Hotels, catering, etc	0.27
25	14	Rubber products	0.26
26	28	Engineers' small tools	0.25
27	13	Processing of plastics	0.23
28	24	Shipbuilding	0.23
29	31	Wooden furniture	0.23
30	23	Motor vehicles	0.21
31	20	Production of steel etc	0.21
32	10	Extraction of mineral oil and gas	0.21
33	34	Printing and publishing	0.21
34	21	Mechanical engineering	0.19
35	41	Construction	0.18
36	42	Trade wholesale & retail	0.16
37	29	Music instruments, toys, etc	0.15
38	37	Clothes	0.15
39	35	Leather, leather goods, footwear	0.14
40	26	Electrical engineering	0.14
41	22	Office machines	0.13
42	27	Instrument engineering	0.12
43	25	Aerospace equipment	0.11
44	47	Other services	0.08
45	45	Banking, finance, insurance, etc	0.08
46	44	Telecommunications	0.06
47	40	Tobacco	0.05

Table 9.4. Ranked CO_2 intensities: Germany 1988.

Rank	Sector	Total Ktonnes CO_2
1	12 Mineral oil processing	155423
2	5 Electricity; distribution	91169
3	11 Chemical products	42383
4	6 Gas	40530
5	38 Food	32578
6	41 Construction	30272
7	42 Trade wholesale & retail	29385
8	23 Motor vehicles	27282
9	43 Traffic & transport services	22587
10	17 Iron and steel, steel products	21431
11	21 Mechanical engineering	18058
12	47 Other services	18056
13	45 Banking, finance, insurance, etc	13710
14	8 Coal extraction, coke ovens, etc	13481
15	26 Electrical engineering	12488
16	46 Hotels, catering, etc	10017
17	1 Agriculture	6215
18	28 Engineers' small tools	5957
19	36 Textile goods	5521
20	31 Wooden furniture	4965
21	32 Pulp, paper, board	4898
22	18 Non-ferrous metals	4492
23	39 Drink	4294
24	15 Stone, clay, cement	3865
25	13 Processing of plastics	3795
26	16 Glass, ceramic goods	3523
27	20 Production of steel etc	3118
28	37 Clothes	2730
29	22 Office machines	2367
30	33 Paper and board products	2008
31	27 Instrument engineering	1837
32	44 Telecommunications	1645
33	14 Rubber products	1515
34	2 Forestry & fishing	1156
35	25 Aerospace equipment	1074
36	29 Music instruments, toys, etc	1038
37	35 Leather, leather goods, footwear	853
38	24 Shipbuilding	830
39	19 Foundries	826
40	34 Printing and publishing	811
41	30 Timber processing	773
42	40 Tobacco	622
43	9 Extraction of metalliferous ores	503
44	7 Water	353
45	10 Extraction of mineral oil and gas	78
46	4 Electricity; other generation	0
47	3 Electricity; fossil generation	0

Table 9.5. Ranked CO_2 emissions: Germany 1988.

9.5.6 CO_2 Emission Elasticities

We have derived all the above elasticities for both Germany and the UK, for all the available input-output tables. We do not present all of these. However, we found that these elasticities were remarkably stable over time, and between countries. This stability was particularly striking with regard to the rankings of the elasticities. As examples of the elasticities, we present some of those for Germany in 1988, in Table 9.6.

Looking first at the column marked 'y', these are the elasticities of CO_2 emission with respect to the elements of final demand. For example, the elasticity for Sector 1 (Agriculture) is 0.014. This shows that if the final demand from the Agriculture sector were to increase by 1%, then overall CO_2 emission would increase by 0.014%. As there are 47 sectors, we would expect that all of these values would be considerably less than unity.

We note that by far the largest final demand elasticity of CO_2 emission is for Electricity; Distribution (5). At 0.202 this is more than twice as large as for the next largest value, which is 0.094 for Chemical Products (11). The next two largest elasticities are for Food (38) and Construction (41).

The next three columns, marked 'gas', 'coal' and 'oil', contain the elasticities of CO_2 emission with the three fossil fuels. They indicate the change of total CO_2 emission for a given change in the output/fuel use coefficients in production; i.e. these elasticities are with respect to the C matrix. The largest elasticity for Gas is for Electricity; Fossil Generation (3). For Coal, there is also the largest elasticity for Electricity; Fossil Generation (3); also large is Coal Extraction (8). For Oil, the largest elasticity is for Traffic and Transport Services (43).

The fifth column is marked 'Col. A', indicating the elasticity of the column sum of the A matrix. This reflects the reduction in CO_2 emissions by the reduction of *all* inputs into a sector. The largest elasticity is for Banking' Finance (45), suggesting that this sector can make the biggest contribution to CO_2 emission reduction by becoming more efficient in its use of inputs.

Column six, marked 'Row A', contains the elasticities of the row sum of the A matrix. This reflects the reduction in CO_2 emissions by the reduction of the use of a sector's output as an input to all sectors. By far the biggest of these elasticities is for Electricity; Distribution (5), indicating that it is by becoming more efficient in the use of electricity *in all sectors* that the greatest reduction in CO_2 emissions is to be achieved.

These elasticities are used in two ways in the remainder of this study. In Chapter 11 they are used to inform the scenario analyses. In Chapter 13 they are a necessary ingredient in the application of the 'minimum disruption' approach to scenario analysis, which is derived in Chapter 12.

168   9 German and UK Input-Output Data for Studying CO_2 Emissions

Sector	y	Production (C) Gas	Coal	Oil	Col A	Row A
1 Agriculture	0.014	0.000	0.000	0.028	0.012	0.007
2 Forestry & fishing	0.003	0.001	0.000	0.004	0.001	0.003
3 Electricity; fossil generation	0.000	0.039	0.352	0.024	0.010	0.157
4 Electricity; other generation	0.000	0.000	0.000	0.000	0.003	0.006
5 Electricity; distribution	0.202	0.000	0.000	0.000	0.026	0.170
6 Gas	0.002	0.000	0.000	0.000	0.002	0.001
7 Water	0.001	0.000	0.000	0.000	0.001	0.002
8 Coal extraction, coke ovens, etc	0.013	0.001	0.141	0.000	0.006	0.093
9 Extraction of metalliferous ores	0.001	0.001	0.000	0.000	0.001	0.006
10 Extraction of mineral oil and gas	0.000	0.000	0.000	0.000	0.000	0.002
11 Chemical products	0.094	0.035	0.012	0.029	0.036	0.021
12 Mineral oil processing	0.009	0.002	0.000	0.008	0.003	0.002
13 Processing of plastics	0.008	0.001	0.000	0.001	0.009	0.005
14 Rubber products	0.003	0.001	0.000	0.001	0.002	0.001
15 Stone, clay, cement	0.009	0.006	0.009	0.007	0.007	0.008
16 Glass, ceramic goods	0.008	0.006	0.000	0.004	0.003	0.005
17 Iron and steel, steel products	0.047	0.016	0.006	0.008	0.035	0.031
18 Non-ferrous metals	0.010	0.003	0.000	0.001	0.006	0.008
19 Foundries	0.002	0.002	0.000	0.001	0.003	0.003
20 Production of steel etc	0.007	0.000	0.000	0.001	0.005	0.002
21 Mechanical engineering	0.040	0.002	0.000	0.006	0.028	0.004
22 Office machines	0.005	0.000	0.000	0.000	0.004	0.000
23 Motor vehicles	0.060	0.004	0.000	0.003	0.043	0.002
24 Shipbuilding	0.002	0.000	0.000	0.000	0.001	0.001
25 Aerospace equipment	0.002	0.000	0.000	0.000	0.002	0.001
26 Electrical engineering	0.028	0.001	0.000	0.005	0.024	0.004
27 Instrument engineering	0.004	0.000	0.000	0.001	0.003	0.000
28 Engineers' small tools	0.013	0.001	0.000	0.002	0.009	0.003
29 Music instruments, toys, etc	0.002	0.000	0.000	0.000	0.001	0.000
30 Timber processing	0.002	0.000	0.000	0.002	0.002	0.004
31 Wooden furniture	0.011	0.000	0.000	0.002	0.007	0.001
32 Pulp, paper, board	0.011	0.004	0.002	0.004	0.004	0.012
33 Paper and board products	0.004	0.001	0.000	0.001	0.005	0.003
34 Printing and publishing	0.002	0.001	0.000	0.001	0.005	0.002
35 Leather, leather goods, footwear	0.002	0.000	0.000	0.000	0.001	0.000
36 Textile goods	0.012	0.003	0.000	0.003	0.006	0.005
37 Clothes	0.006	0.000	0.000	0.001	0.005	0.000
38 Food	0.072	0.009	0.001	0.013	0.040	0.006
39 Drink	0.010	0.002	0.000	0.003	0.006	0.002
40 Tobacco	0.001	0.000	0.000	0.000	0.001	0.000
41 Construction	0.067	0.000	0.000	0.017	0.038	0.003
42 Trade wholesale & retail	0.065	0.005	0.000	0.029	0.032	0.010
43 Traffic & transport services	0.050	0.000	0.000	0.072	0.018	0.018
44 Telecommunications	0.004	0.000	0.000	0.002	0.001	0.001
45 Banking, finance, insurance, etc	0.030	0.001	0.000	0.003	0.059	0.002
46 Hotels, catering, etc	0.022	0.002	0.000	0.005	0.012	0.002
47 Other services	0.040	0.004	0.000	0.025	0.060	0.012

Table 9.6. CO_2 emissions: elasticities with respect to final demand, fuels, energy, column sums and row sums: Germany 1988.

9.6 Conclusions

Using published input-output tables, and other available energy statistics, we have implemented the theory on input-output analysis of CO_2 emissions, developed in Chapter 8.

From the general discussion it is apparent that the electricity industry will be of central importance in any attempt at reducing CO_2 emissions. However, the data so far developed does not allow for the effects of imports and exports, nor for the dynamics of CO_2 emissions, as related to economic structural change. It is to this area we turn in the next chapter.

10 Input-Output Analysis of German and UK CO_2 Emissions

10.1 Introduction

In Chapter 9 we outlined the input-ouput data used in this study, and made some preliminary analyses. In this chapter we shall continue our input-output analysis of German and UK CO_2 emissions, using the theory established in Chapter 8. The remainder of this chapter has the following structure.

In Section 10.2 we discuss the attribution of CO_2 emissions, in particular between domestic consumption and exports. Section 10.3 extends this discussion, to take account of imports. Here we distinguish between CO_2 emissions and CO_2 'responsibility'. The decomposition of changing CO_2 emissions over time is discussed in Section 10.4, and the decomposition of the differences in CO_2 emissions between Germany and the UK is made in Section 10.5. In Section 10.6 we summarise our conclusions.

10.2 Attribution of CO_2 Emissions by Germany and the UK

We begin by examining the time-trends of the various components of total CO_2 emissions by Germany and the UK. We recall that in Section 8.4.1 we derived:

$$C = \mathbf{e'P'Z}y^d + (\mathbf{e'C'} + \mathbf{m'})(\mathbf{I} - \mathbf{A})^{-1}y^d \qquad (8.23)$$
$$+ (\mathbf{e'C'} + \mathbf{m'})(\mathbf{I} - \mathbf{A})^{-1}y^e.$$

Here we distinguish between final demand consumed domestically (y^d) and final demand for export (y^e).

We also recall that we can expand the Leontief inverse matrix as an infinite series, as shown in Section 7.5:

$$(\mathbf{I} - \mathbf{A})^{-1} = \mathbf{I} + \mathbf{A} + \mathbf{A}^2 + \mathbf{A}^3 + \ldots \qquad (7.33)$$

10.2 Attribution of CO_2 Emissions by Germany and the UK

Combining (7.33) and (8.23), and reorganising, gives:

$$C = e'P'Zy^d$$
$$+ (e'C' + m')y^d$$
$$+ (e'C' + m')(A + A^2 + A^3 ...)y^d \qquad (10.1)$$
$$+ (e'C' + m')(I - A)^{-1}y^e.$$

We can interpret equation (10.1) as follows. The first part of the right-hand side represents the CO_2 emission attributable directly to the final demand for fuels. The second part corresponds to the CO_2 emission attributable to *direct* fossil fuel use for producing goods for domestic final demand. The third part corresponds to the CO_2 emission attributable to *indirect* fossil fuel use for producing goods for domestic final demand. The fourth part corresponds to CO_2 emission attributable to the production of final demand for export.

From the input-output data discussed in Chapter 9 we can derive these four components, for both Germany and the UK, for each available input-output table. We present this decomposition of CO_2 emissions in both absolute and proportional terms, in Table 10.1.

From the table it is clear that the four components are relatively stable over time, for both Germany and the UK. Further, the components have very similar relative sizes for the two countries. This is revealed most clearly by plotting the relative size components individually, for both Germany and the UK, as in Figures 10.1 to 10.5.

We look first at Figure 10.1, which shows the proportions of CO_2 emissions attributable directly to household use of fossil fuels. It is noticeable that both the UK (1968-1984) and Germany (1978-1988) show upward trends in the proportions of total CO_2 emission attributable to household fuel use until 1986. It is also noticeable that the time-trends for the two countries are almost coincident. In 1968 only 21% of UK CO_2 emission is directly attributable to household fossil fuel use (though this excludes electricity use); by 1984 this is about 30% for both Germany and the UK. This rapid rise can probably be attributed to improvements in the standard of home heating, and the very rapid increase in the ownership and use of motor cars. Both correspond to increasing Gross Domestic Product (GDP) per capita, as discussed in Chapter 5.

10 Input-Output Analysis of German and UK CO_2 Emissions

Country	Date	Final Demand	Direct Manuf'ing	Indirect Manuf'ing	Domestic Total	Exports	CO_2 Emission
Germany	1978	183165.4	71183.0	293002.8	547351.3	164045.5	711396.7
	%	25.7	10.0	41.2	76.9	23.1	100.0
Germany	1980	197946.3	71348.5	317705.9	587000.6	173144.2	760144.8
	%	26.0	9.4	41.8	77.2	22.8	100.0
Germany	1982	181748.7	63818.8	288389.3	533956.8	163109.1	697065.9
	%	26.1	9.2	41.4	76.6	23.4	100.0
Germany	1984	192033.4	57229.3	262380.7	511643.5	153338.7	664982.2
	%	28.9	8.6	39.5	76.9	23.1	100.0
Germany	1986	208896.0	63268.9	262580.4	534745.4	147042.3	681787.7
	%	30.7	9.3	38.5	78.4	21.6	100.0
Germany	1988	198679.0	61261.0	250072.4	510012.4	140528.6	650541.0
	%	30.5	9.5	38.4	78.4	21.6	100.0
UK	1968	129185.8	88967.7	294017.6	512171.2	100605.1	612776.2
	%	21.1	14.5	48.0	83.6	16.4	100.0
UK	1974	153429.5	54655.8	285426.7	493512.0	133381.6	626893.6
	%	24.5	8.7	45.5	78.7	21.3	100.0
UK	1979	168498.1	54067.7	274417.1	496983.0	148601.8	645584.8
	%	26.1	8.4	42.5	77.0	23.0	100.0
UK	1984	161041.6	35376.1	218251.9	414669.6	111251.3	525920.8
	%	30.6	6.7	41.5	78.8	21.2	100.0

Table 10.1. Emissions of CO_2 to satisfy domestic and export final demand (Ktonnes): Germany and the UK.[1]

Turning now to Figure 10.2, the share of direct manufacturing to satisfy domestic final demand has been falling for the UK, from over 14% in 1968, to less than 7% by 1984. For Germany there has also been a slight decline, from rather more than 10% in 1978, to less than 10% in 1988. This fall cannot be attributed to one particular factor, but in part must reflect growing efficiency in fuel use in production.

In Figure 10.3 we see the share of indirect manufacturing CO_2 emissions falls for both Germany and the UK. We also note that the trends for the two countries are almost coincident. This fall probably reflects two trends in the manufacturing sector. First, increasing efficiency of fuel use; second, a *decreasing* degree of interconnectedness in both economies. (This issue will be taken up again in Section 10.4 below.)

---

[1] The fall in actual CO_2 emissions for the UK, between 1979 and 1984 is attributable to the major coal-miners' strike during the latter year, which sharply reduced coal consumption in electricity production.

10.2 Attribution of CO_2 Emissions by Germany and the UK

The overall domestic shares of CO_2 emission, for Germany and the UK, are shown in Figure 10.4. This is simply the sum of the figures displayed in Figures 10.1 to 10.3. We see that for the UK, the share of CO_2 emissions attributable to domestic demand for fuels and goods by households falls from over 83% in 1968, to 79% by 1984. In Germany, the domestic share of total CO_2 emissions actually rises, from 77% in 1978 to over 78% by 1988. We again note that during the period of overlap of the data, the proportions for Germany and the UK are extremely close.

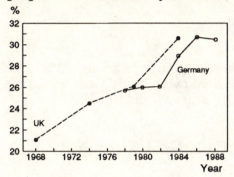

Figure 10.1. CO_2 emission by Germany and the UK: domestic final demand proportion.

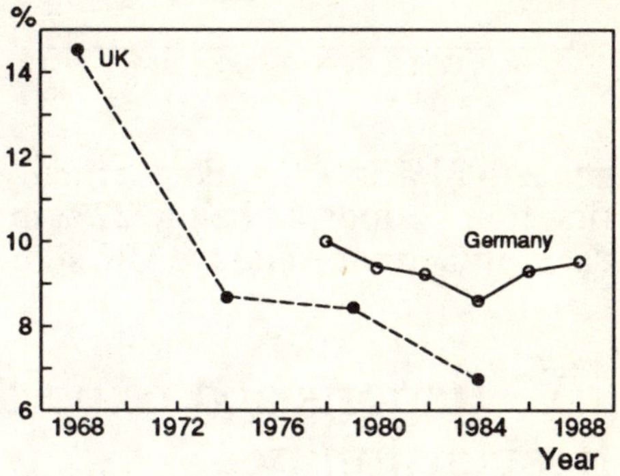

Figure 10.2. CO_2 emission by Germany and the UK: domestic direct manufacturing proportion.

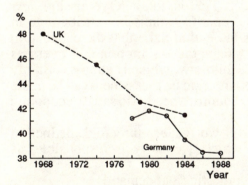

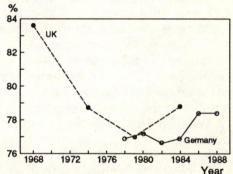

Figure 10.3. CO_2 emission by Germany and the UK: domestic indirect manufacturing proportion.

Figure 10.4. CO_2 emission by Germany and the UK: domestic total proportion.

Finally, in Figure 10.5 we see the share of total CO_2 emissions by Germany and the UK which can be attributed to the final demand for exports. For the UK the share rises from nearly 17% in 1968, to 23% in 1979, declining

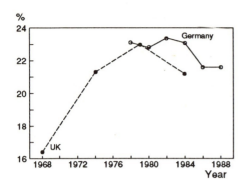

Figure 10.5. CO_2 emission by Germany and the UK: exports proportion.

again to 21% in 1984. For Germany, there is a fall from around 23% in the first four periods, to below 22% in the final two periods. Once again, the data for Germany and the UK show similar levels and trends.

10.3 Imports and Exports: CO_2 Emission and Responsibility

In Section 8.4 we showed how the input-output approach could be extended to include imports and exports. This allows us to calculate the total CO_2 responsibility of Germany and the UK. That is, total CO_2 emission, *less* CO_2 emission attributable to exports, *plus* CO_2 emission attributable to imports. We recall from the analysis in Section 8.4.3 that the calculation of CO_2 attributable to imports necessitates the assumption that overseas technology is the same as domestic technology, which is almost certainly not the case. However, without this assumption no assessment can be made of the amount of CO_2 emission in other countries needed to satisfy economic activity in the domestic country.

We also recall that in Section 8.4.3 we made a threefold distinction concerning the CO_2 emissions for which a country is responsible, as follows:

Domestic CO_2 emission to meet Domestic final demand
$$= \mathbf{r}^{1\prime}(\mathbf{I} - \mathbf{A}_1)^{-1}\mathbf{y}^{11} \qquad (8.52)$$

Foreign CO_2 emission to meet Domestic final demand
$$= \mathbf{r}^{1\prime}(\mathbf{I} - \mathbf{A}_1)^{-1}\mathbf{B}_2(\mathbf{I} - \mathbf{A}_1)^{-1}\mathbf{y}^{11} \qquad (8.53)$$

10.3 Imports and Exports: CO_2 Emission and Responsibility

Foreign CO_2 emission to meet Imported final demand
$$= \mathbf{r}^{1\prime}(\mathbf{I}-\mathbf{A}_1)^{-1}\mathbf{y}^{21}. \qquad (8.54)$$

For brevity, we refer to these elements as 'Domestic', 'Direct Imports' and 'Indirect Imports', respectively.

This decomposition allows us to construct Table 10.2, where the absolute amounts of CO_2 emission are shown, for both countries, over all periods. The decomposition is into the Domestic Total CO_2 emission, the Direct, Indirect and Total Imports. These sum to total CO_2 emission responsibility. Also shown are the proportions of CO_2 emission *responsibility*. In the final column is displayed the ratio of CO_2 responsibility to total CO_2 emission.

Country	Date	Domestic Total	Direct Imports	Indirect Imports	Total Imports	CO_2 Responsibility	% Respons/ Emission
Germany	1978	547351.3	32283.2	69815.8	102099.0	649450.2	91.3
%		84.3	5.0	10.7	15.7	100.0	
Germany	1980	587000.6	38282.1	83146.8	121428.9	708429.6	93.2
%		82.9	5.4	11.7	17.1	100.0	
Germany	1982	533956.8	32535.3	74625.2	107160.5	641117.3	92.0
%		83.3	5.1	11.6	16.7	100.0	
Germany	1984	511643.5	34679.7	69602.3	104282.0	615925.5	92.6
%		83.1	5.6	11.3	16.9	100.0	
Germany	1986	534745.4	42614.5	62285.1	104899.5	639644.9	93.8
%		83.6	6.7	9.7	16.4	100.0	
Germany	1988	510012.4	36074.5	58992.1	95066.6	605078.9	93.0
%		84.3	6.0	9.7	15.7	100.0	
UK	1968	512171.1	24053.4	49982.9	74036.3	586207.4	95.7
%		87.4	4.1	8.5	12.6	100.0	
UK	1974	493512.0	38293.4	59565.2	97858.6	591370.6	94.3
%		83.5	6.5	10.1	16.5	100.0	
UK	1979	496983.0	42959.8	67522.4	110482.4	607465.4	94.1
%		81.8	7.1	11.1	18.2	100.0	
UK	1984	414669.6	45403.3	49145.9	94549.2	509218.8	96.8
%		81.4	8.9	9.7	18.6	100.0	

Table 10.2. Responsibilities for CO_2 Emissions (K tonnes): Germany and the UK.

From Table 10.2 we can plot the proportional figures for the various components, as in Figures 10.6 to 10.9.

Considering Figure 10.6, we see that the proportion of domestic CO_2 emission in total CO_2 responsibility has been falling for the UK, from over 87% in 1969 to less than 82% in 1984. For Germany this figure has fluctuated between 83% and 84.5%.

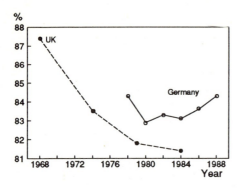
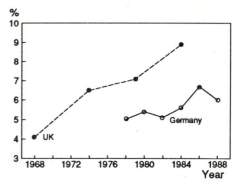

Figure 10.6. CO_2 responsibility by Germany and the UK: domestic proportion.

Figure 10.7. CO_2 responsibility by Germany and the UK: direct imports proportion.

The proportion of responsibility of CO_2 attributable to direct imports is shown in Figure 10.5. We see that for the UK this has steadily increased, from 4% in 1969 to 9% in 1984. There has also been an increase for Germany, from 5% in 1978 to a peak of nearly 7% in 1986. It is interesting to consider why the UK proportion has grown so rapidly compared with Germany. This can probably be put down to the shift in the UK away from industrial manufacturing, and towards the provision of services. This has entailed the import of industrially manufactured goods as a growing proportion of GDP, which will have caused this CO_2 responsibility to increase.

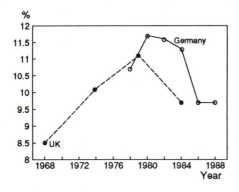
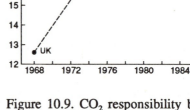

Figure 10.8. CO_2 responsibility by Germany and the UK: indirect imports proportion.

Figure 10.9. CO_2 responsibility by Germany and the UK: total imports proportion.

In Figure 10.8 is shown the proportion of indirect imports in CO_2 responsibility. For the UK this also increases, from 8.5% to 11%, except between the final two periods. For Germany the trend is generally downwards, from a peak Of 11.5% to 10%. These trends can probably be attributed to the increasing tendency in the UK to meet final demand directly by imports.

Figure 10.9 shows the proportion of CO_2 responsibility accounted for by total imports. The UK shows a clear upward trend, from 12.5% in 1969, to 18.5% in 1984. Germany, on the other hand, shows a reasonably steady proportion, of about 16.5%.

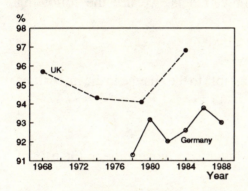

Figure 10.10. Ratio of CO_2 responsibility to emissions: Germany and the UK.

Finally, in Figure 10.10 we see the ratio of CO_2 responsibility to CO_2 emission by Germany and the UK. The trend for the UK was downward until 1979. The marked increase in that year can probably be attributed to the year-long coal miners' strike, with the distortions to trade in both fuel and non-fuel commodities that it caused. However, the ratio is clearly less than unity for the UK, indicating that, on the basis of the assumptions used, the UK is *responsible* for less CO_2 than it emits. For Germany the proportion has increased from 91% to about 93.5%, but is also clearly less than unity. Germany is also *responsible* for less CO_2 than it actually emits, when trade patterns are taken into account.

10.4 Changes in CO_2 Emission over Time

We now turn to an analysis of how CO_2 emissions have changed over time, in Germany and the UK. In particular we shall apply the decomposition by differencing discussed in Section 8.3. We recall that this gave rise to:

$$\Delta C = \mathbf{e}'[\Delta \mathbf{P}'\overline{\mathbf{Z}\mathbf{u}\mathbf{Y}} + \overline{\mathbf{P}}'\Delta \mathbf{Z}\overline{\mathbf{u}\mathbf{Y}} + \overline{\mathbf{P}}'\overline{\mathbf{Z}}\Delta \mathbf{u}\overline{\mathbf{Y}} + \overline{\mathbf{P}}'\overline{\mathbf{Z}\mathbf{u}}\Delta \mathbf{Y}]$$
$$+ [(\mathbf{e}'\Delta \mathbf{C}' + \Delta \mathbf{m}')\overline{(\mathbf{I}-\mathbf{A})^{-1}\mathbf{u}\mathbf{Y}} + (\mathbf{e}'\overline{\mathbf{C}} + \overline{\mathbf{m}}')\Delta(\mathbf{I}-\mathbf{A})^{-1}\overline{\mathbf{u}\mathbf{Y}} \quad (8.19)$$
$$+ (\mathbf{e}'\overline{\mathbf{C}}' + \overline{\mathbf{m}}')\overline{(\mathbf{I}-\mathbf{A})^{-1}}\Delta \mathbf{u}\overline{\mathbf{Y}} + (\mathbf{e}'\overline{\mathbf{C}}' + \overline{\mathbf{m}}')\overline{(\mathbf{I}-\mathbf{A})^{-1}\mathbf{u}}\Delta \mathbf{Y}]$$
$$+ remainder.$$

In the tables representing these eight elements, we use the following 'shorthand' notation.

1. 'ΔP' $\Leftrightarrow \mathbf{e}'\Delta \mathbf{P}'\overline{\mathbf{Z}\mathbf{u}\mathbf{Y}}$.

   The change in CO_2 emissions attributable to the change in the intensities of fuel use by final demand.

2. 'ΔZ' $\Leftrightarrow \mathbf{e}'\overline{\mathbf{P}}'\Delta \mathbf{Z}\overline{\mathbf{u}\mathbf{Y}}$.

   The change in CO_2 emissions attributable to the change in the mix of final demand type for fuels.

3. 'Δu_P' $\Leftrightarrow \mathbf{e}'\overline{\mathbf{P}}'\overline{\mathbf{Z}}\Delta \mathbf{u}\overline{\mathbf{Y}}$.

   The change in CO_2 emissions attributable to the change in the relative sectoral shares of fuel demand by final demand.

4. 'ΔY_P' $\Leftrightarrow \mathbf{e}'\overline{\mathbf{P}}'\overline{\mathbf{Z}\mathbf{u}}\Delta \mathbf{Y}$.

   The change in CO_2 emissions attributable to the change in overall domestic purchasing power of final demand for fuels.

5. 'ΔC' $\Leftrightarrow (\mathbf{e}'\Delta \mathbf{C}' + \Delta \mathbf{m}')\overline{(\mathbf{I}-\mathbf{A})^{-1}\mathbf{u}\mathbf{Y}}$.

   The change in CO_2 emissions attributable to the change in the intensities of fuel use in production.

6. 'ΔL' $\Leftrightarrow (\mathbf{e}'\overline{\mathbf{C}} + \overline{\mathbf{m}}')\Delta(\mathbf{I}-\mathbf{A})^{-1}\overline{\mathbf{u}\mathbf{Y}}$.

   The change in CO_2 emissions attributable to the change in the structure of interindustry trading.

7. 'Δu_C' $\Leftrightarrow (\mathbf{e}'\overline{\mathbf{C}}' + \overline{\mathbf{m}}')\overline{(\mathbf{I}-\mathbf{A})^{-1}}\Delta \mathbf{u}\overline{\mathbf{Y}}$.

The change in CO_2 emissions attributable to the change in the final demand proportions for the goods from the various sectors.

8. 'ΔY_C' $\Leftrightarrow (e'\overline{C}' + \overline{m}')(I-A)^{-1}\overline{u}\Delta Y$.

The change in CO_2 emissions attributable to the change in the overall purchasing power for produced goods.

Elements 1-4 refer only to the final demand use of fuels, so make reference only to sectors 6 (gas), 8(coal) and 12 (oil).

Using the 'short-hand' notation above, we can write equation (8.19) as:

$$\Delta C = \Delta P + \Delta Z + \Delta u_p + \Delta Y_p \\ + \Delta C + \Delta L + \Delta u_C + \Delta Y_C. \quad (10.2)$$

Now equation (8.20) can be represented in two ways. First, as a full decomposition by each sector. Second, as a decomposition of the total components, summed over all sectors.

10.4.1 The Full Sectoral Decomposition

As representatives of the full decomposition, in Tables 10.3 and 10.4 we show the decompositions for Germany (1986-88) and the UK (1979-84) respectively. (Complete tables for the decomposition of all adjacent I-O tables, for both Germany and the UK, are given in Appendix A10.) The figures are shown as annualised percentage rates of change.

The first three rows, marked 'Final Demand' represent the decomposition of the change in CO_2 emissions attributable directly to the final demand for fuels. These represent only the contributions by the three sectors: Gas (6); Coal extraction, coke ovens, etc (8); Mineral oil processing (12). This is because fossil fuels are bought by households directly from these three sectors. These rows are subdivided into four columns, representing elements 1-4 described above. The fifth column shows the sum of the preceding four columns.

The fourth row, marked 'Total', shows the overall effect of changing use of these fuels for final demand, upon total final demand CO_2 emissions. This is also subdivided into four columns.

The succeeding 47 rows, marked 'Production', represent the decomposition of the change in CO_2 emissions attributable to production. These are subdivided into four columns, representing elements 5-8 of the decomposition described above. The fifth column is the sum of the preceding four columns. Column six is the sum of columns five for final demand and production. It differs from column five for production only in the fossil fuel sectors (6, 8, 12).

The final row shows the total components of changes in CO_2 emissions, summed over all sectors.

For example, for Germany 1986-88 we see that for Agriculture, the CO_2 emission increased by 2.05%. This is accounted for as follows.

As the Agriculture sector is not a fuel sector, there are no contributions to CO_2 emissions attributable to direct consumption of fuels by households.

The 'ΔC' element is -4.59% p.a. That is, agricultural CO_2 emission reduced at this rate, because of changing fuel mix and improving energy efficiency in the agricultural sector.

The '$\Delta(I-A)^{-1}$' element is -0.34% p.a., indicating that because of changing interindustry trading relations, CO_2 emissions attributable to agriculture reduced at this rate.

The '$\Delta \mathbf{u}$' element is 2.67% p.a. This is the proportional rate of increase in agricultural CO_2 emission attributable to the changing mix of final demand in the economy.

The 'ΔY' element is 4.31% p.a., and reflects the effect of aggregate growth in GDP on the aggregate demand for agricultural products.

The 'Sum' element is the total of the above four elements, and is 2.05% p.a. Thus the overall effect of changes in fuel use, interindustry trading, and the structure and level of GDP caused overall this annualised rate of increase of CO_2 emissions attributable to Agriculture.

The sectors Gas, Coal Extraction and Mineral Oil Processing are represented in the first three rows. This reflects these sectors selling fuels directly to final demand, so that some CO_2 emissions should be attributed to this consumption activity. Looking at the Gas sector, we can examine each of these elements in turn.

10.4.1 The Full Sectoral Decomposition

Final Demand (% p.a.)	ΔP	ΔZ	Δu_P	ΔY_P	Sum	
6 Gas	-10.74	0.04	7.41	4.20	0.91	
8 Coal extraction, coke ovens, etc	21.54	-14.23	-19.99	2.27	-10.42	
12 Mineral oil processing	3.19	-1.88	-7.68	4.11	-2.26	
Total	0.79	-0.89	-2.02	1.31	-0.81	

Production (% p.a.)	ΔC	ΔL	Δu_C	ΔY_C	Sum	Total
1 Agriculture	-4.59	-0.34	2.67	4.31	2.05	2.05
2 Forestry & fishing	-6.23	0.20	-1.35	4.21	-3.18	-3.18
3 Electricity; fossil generation	0.00	0.00	0.00	0.00	0.00	0.00
4 Electricity; other generation	0.00	0.00	0.00	0.00	0.00	0.00
5 Electricity; distribution	-5.70	-0.61	-2.60	4.17	-4.74	-4.74
6 Gas	-0.18	-0.48	0.19	0.11	-0.36	0.55
7 Water	-5.49	-3.45	-4.34	4.07	-9.21	-9.21
8 Coal extraction, coke ovens, etc	-1.87	-0.30	-17.14	1.95	-17.36	-27.78
9 Extraction of metalliferous ores	-13.06	2.89	3.72	4.26	-2.20	-2.20
10 Extraction of mineral oil and gas	-5.52	12.00	10.50	4.74	21.72	21.72
11 Chemical products	-1.83	-1.86	3.14	4.35	3.81	3.81
12 Mineral oil processing	-0.14	-0.01	-0.21	0.11	-0.25	-2.51
13 Processing of plastics	-4.76	0.53	3.74	4.36	3.87	3.87
14 Rubber products	-7.48	-1.89	2.28	4.22	-2.87	-2.87
15 Stone, clay, cement	-8.47	-0.42	1.23	4.21	-3.45	-3.45
16 Glass, ceramic goods	-12.88	0.75	1.12	4.15	-6.87	-6.87
17 Iron and steel, steel products	-1.74	-2.81	2.84	4.33	2.61	2.61
18 Non-ferrous metals	-6.27	4.06	-7.30	4.17	-5.33	-5.33
19 Foundries	-4.26	2.34	2.27	4.37	4.73	4.73
20 Production of steel etc	-6.18	-0.34	0.55	4.24	-1.73	-1.73
21 Mechanical engineering	-4.31	-1.64	-3.09	4.17	-4.88	-4.88
22 Office machines	-6.19	1.51	-4.87	4.16	-5.38	-5.38
23 Motor vehicles	-5.73	2.08	-0.74	4.27	-0.12	-0.12
24 Shipbuilding	-4.25	1.23	-10.70	4.08	-9.64	-9.64
25 Aerospace equipment	-4.97	-4.06	1.25	4.20	-3.59	-3.59
26 Electrical engineering	-6.27	-0.03	1.90	4.27	-0.12	-0.12
27 Instrument engineering	-5.96	-1.69	1.76	4.24	-1.64	-1.64
28 Engineers' small tools	-5.17	-0.93	1.90	4.28	0.08	0.08
29 Music instruments, toys, etc	-5.43	2.40	-0.75	4.29	0.50	0.50
30 Timber processing	-7.98	-0.52	6.01	4.33	1.83	1.83
31 Wooden furniture	-6.80	0.76	0.30	4.24	-1.50	-1.50
32 Pulp, paper, board	-7.29	1.59	7.22	4.41	5.92	5.92
33 Paper and board products	-5.00	0.04	1.56	4.29	0.89	0.89
34 Printing and publishing	-5.77	3.89	-1.82	4.29	0.59	0.59
35 Leather, leather goods, footwear	-5.13	3.11	-3.63	4.25	-1.40	-1.40
36 Textile goods	-4.42	-0.15	-1.08	4.24	-1.41	-1.41
37 Clothes	-5.13	1.45	-3.53	4.21	-3.01	-3.01
38 Food	-3.94	2.56	-3.20	4.27	-0.31	-0.31
39 Drink	-6.99	-0.05	-0.73	4.20	-3.56	-3.56
40 Tobacco	-6.07	0.90	-3.77	4.18	-4.77	-4.77
41 Construction	-6.79	-0.94	-0.39	4.19	-3.93	-3.93
42 Trade wholesale & retail	-5.29	-0.54	-0.42	4.23	-2.02	-2.02
43 Traffic & transport services	-12.85	6.42	6.93	4.40	4.90	4.90
44 Telecommunications	-5.50	3.55	1.46	4.37	3.87	3.87
45 Banking, finance, insurance, etc	-6.26	0.28	-0.69	4.22	-2.45	-2.45
46 Hotels, catering, etc	-5.95	-0.31	0.09	4.23	-1.94	-1.94
47 Other services	-6.14	-2.53	2.76	4.24	-1.66	-1.66
Total	-3.80	-0.04	-0.65	2.93	-1.56	-2.37

Table 10.3. Attribution of difference in CO_2 emission between Germany 1986 and Germany 1988 (p.a.).

182   10 Input-Output Analysis of German and UK CO_2 Emissions

Final Demand (% p.a.)	ΔP	ΔZ	Δu_P	ΔY_P	Sum	
6 Gas	0.80	0.06	1.07	0.68	2.61	
8 Coal extraction, coke ovens, etc	1.13	7.26	-17.23	0.69	-8.15	
12 Mineral oil processing	5.99	-0.70	-6.57	0.62	-0.67	
Total	1.11	0.23	-1.78	0.20	-0.24	
Production (% p.a.)	ΔC	ΔL	Δu_C	ΔY_C	Sum	Total
1 Agriculture	-5.50	-3.81	9.46	0.73	0.88	0.88
2 Forestry & fishing	-12.53	3.16	0.04	0.68	-8.65	-8.65
3 Electricity; fossil generation	0.00	0.00	0.00	0.00	0.00	0.00
4 Electricity; other generation	0.00	0.00	0.00	0.00	0.00	0.00
5 Electricity; distribution	-4.21	-0.27	-1.95	0.68	-5.75	-5.75
6 Gas	0.04	-0.06	0.03	0.02	0.03	2.64
7 Water	-1.10	-3.40	0.91	0.68	-2.91	-2.91
8 Coal extraction, coke ovens, etc	0.39	1.30	-1.48	0.06	0.27	-7.88
9 Extraction of metalliferous ores	7.27	0.86	-42.13	0.88	-33.13	-33.13
10 Extraction of mineral oil and gas	-3.83	13.16	11.56	0.80	21.69	21.69
11 Chemical products	-4.34	-3.33	3.36	0.69	-3.62	-3.62
12 Mineral oil processing	-0.50	0.37	-0.95	0.09	-0.99	-1.65
13 Processing of plastics	1.14	-3.88	-3.12	0.67	-5.19	-5.19
14 Rubber products	3.17	-3.76	-3.96	0.68	-3.88	-3.88
15 Stone, clay, cement	-4.75	-0.08	-6.92	0.66	-11.09	-11.09
16 Glass, ceramic goods	-6.20	0.05	-0.79	0.68	-6.26	-6.26
17 Iron and steel, steel products	-7.61	-2.09	-1.87	0.67	-10.90	-10.90
18 Non-ferrous metals	-0.04	1.46	5.88	0.73	8.03	8.03
19 Foundries	-5.80	-0.87	-6.29	0.66	-12.31	-12.31
20 Production of steel etc	-5.10	-0.90	-4.97	0.66	-10.31	-10.31
21 Mechanical engineering	-8.70	-0.82	-8.05	0.65	-16.92	-16.92
22 Office machines	-4.93	-7.99	6.84	0.71	-5.37	-5.37
23 Motor vehicles	-4.76	-4.59	-4.08	0.65	-12.78	-12.78
24 Shipbuilding	-6.62	-0.56	-1.09	0.68	-7.59	-7.59
25 Aerospace equipment	-7.53	-3.38	6.30	0.71	-3.90	-3.90
26 Electrical engineering	-4.68	-4.77	4.62	0.69	-4.15	-4.15
27 Instrument engineering	-3.74	-5.00	-4.06	0.65	-12.15	-12.15
28 Engineers' small tools	-4.57	0.92	9.85	0.74	6.94	6.94
29 Music instruments, toys, etc	0.07	-0.55	-11.15	0.67	-10.97	-10.97
30 Timber processing	2.85	-0.41	-6.82	0.70	-3.68	-3.68
31 Wooden furniture	2.24	-1.35	0.53	0.70	2.12	2.12
32 Pulp, paper, board	-8.35	-0.26	1.73	0.69	-6.19	-6.19
33 Paper and board products	-4.86	-1.00	3.53	0.69	-1.64	-1.64
34 Printing and publishing	-5.03	1.17	-2.96	0.68	-6.14	-6.14
35 Leather, leather goods, footwear	-8.17	-2.00	-1.49	0.66	-11.00	-11.00
36 Textile goods	-9.99	-4.90	-2.11	0.66	-16.35	-16.35
37 Clothes	-8.99	-5.69	-3.83	0.64	-17.86	-17.86
38 Food	-4.60	-1.28	-0.85	0.67	-6.06	-6.06
39 Drink	-5.73	1.04	-3.96	0.67	-7.99	-7.99
40 Tobacco	-7.70	5.29	-9.11	0.68	-10.84	-10.84
41 Construction	-3.67	1.67	1.65	0.70	0.36	0.36
42 Trade wholesale & retail	-5.66	2.46	1.86	0.70	-0.64	-0.64
43 Traffic & transport services	0.74	1.47	-4.31	0.69	-1.41	-1.41
44 Telecommunications	-5.20	8.49	-2.16	0.72	1.86	1.86
45 Banking, finance, insurance, etc	-5.44	7.75	0.98	0.72	4.00	4.00
46 Hotels, catering, etc	-6.98	-4.96	2.27	0.68	-8.99	-8.99
47 Other services	-4.99	12.77	-0.88	0.75	7.65	7.65
Total	-3.19	-0.19	-0.89	0.49	-3.77	-4.02

Table 10.4. Attribution of difference in CO_2 emission between UK 1979 and UK 1984 (% p.a.).

The 'ΔP' element is -10.74% p.a. This reflects the rate of reduction in CO_2 emissions attributable to the altering relationship of the quantity of gas purchased for final demand to the amount spent. That is, it reflects a falling gas price.

The 'ΔZ' element is 0.04% p.a. This reflects the rate of alteration of the proportion of final demand for gas used domestically.

The 'Δu' element is 7.41% p.a., which reflects the increasing share of expenditure on gas purchases by final demand.

The 'ΔY' element is 4.20, which reflects the effect of aggregate economic growth on the final demand for gas.

The 'Sum' element is 0.91% p.a., indicating that the overall effect of altering the final demand use of gas has been to increase attributable CO_2 emissions at this rate.

The elements corresponding to manufacturing gas use then follow, and these sum to -0.36% p.a. Finally, there is the 'Total' element of 0.55% p.a. This is the sum of the eight individual elements (or the two 'Sum' elements). It shows that in 1986-88 total CO_2 emission attributable to the German Gas industry increased at this annualised rate.

There is also a 'remainder' term involved in this decomposition, as discussed in Chapters 4 and 8. However, this is generally very small, so for the sake of brevity it has been omitted.

10.4.2 The Aggregate Decomposition

The sum of all the sectoral elements can also be represented as the proportional change of *total* CO_2 emissions from the base year figure. These elements have been combined from the individual tables in the above section, and Appendix A10, and are displayed in Table 10.5. They provide a concise summary of how CO_2 emissions have changed over time, as attributed to the various factors.

For example, for Germany 1978-80, the total 'ΔP' element is -1.01% % p.a., reflecting the effect of changing fuel prices on CO_2 emissions attributable to direct final demand for fuels.

The 'ΔZ' element is 1.35% p.a., indicating the effect of the changing proportion of domestic demand in total final demand for fuels.

The first 'Δu' element is -0.18% p.a., and reflects the effect of the changing mix of final demand expenditure on fuels. The fact that this is negative reflects a shift from coal towards the less carbon intensive oil and gas.

10 Input-Output Analysis of German and UK CO_2 Emissions

Country	Dates	ΔP	ΔZ	Δu_p	ΔY_p	ΣP	ΔC	ΔL	Δu_c	ΔY_c	ΣC	$\Sigma P+C$
Germany	1978-80	-1.01	1.35	-0.18	0.75	0.91	0.88	0.31	-0.89	2.04	2.35	3.26
Germany	1980-82	0.72	0.30	-2.04	-0.06	-1.09	0.82	-3.85	0.02	-0.18	-3.19	-4.28
Germany	1982-84	0.42	1.10	-1.36	0.58	0.75	-3.45	-0.69	-0.47	1.53	-3.08	-2.33
Germany	1984-86	0.32	0.10	0.20	0.64	1.26	1.98	-3.23	-0.27	1.51	0.00	1.26
Germany	1986-88	0.79	-0.89	-2.02	1.31	-0.81	-3.80	-0.04	-0.65	2.93	-1.56	-2.37
UK	1968-74	-1.72	-0.23	1.94	0.71	0.69	-3.83	0.84	0.65	2.12	-0.30	0.39
UK	1974-79	-0.73	0.57	-0.03	0.69	0.50	-0.13	-1.59	-0.18	2.02	0.13	0.63
UK	1979-84	1.11	0.23	-1.78	0.20	-0.24	-3.19	-0.19	-0.89	0.49	-3.77	-4.02

Table 10.5. CO_2 emission changes (% p.a.): Germany and the UK.

The first 'ΔY' element is 0.75% % p.a., which is the rate of increase in CO_2 emissions attributable to the increase in the purchasing of fuels because, of increased overall purchasing power.

The first 'Sum' element is 0.91% p.a., which is the total annual rate of increase of CO_2 emissions attributable to changed fuel purchasing by households.

The 'ΔC' element is 0.88% p.a., and reflects the increase in CO_2 emissions attributable to overall changes in fossil fuel mix and efficiency of use, in the production part of the economy.

The '$\Delta(I-A)^{-1}$' element is 0.31% p.a., indicating that altering industry trading relations in that period led to an increase in overall CO_2 emissions in Germany.

The second 'Δu' term is -0.89% p.a. This is the rate of reduction in CO_2 emissions attributable to the changing mix of final demand for all goods in Germany.

The second 'ΔY' term is 2.04% p.a., which is the rate of increase in German CO_2 emissions during this period, attributable to growth in aggregate final demand.

Now in this study we are particularly concerned to understand how CO_2 emissions change over time, and how they differ between Germany and the UK. We therefore present plots of the various components of Table 10.5, in Figures 10.11 to 10.21. Here we plot the calculated differences at the mid-period for the corresponding difference. For example, the UK 1968-74 data are plotted at 1971; the German 1978-80 data at 1979, etc.

10.4.2 The Aggregate Decomposition

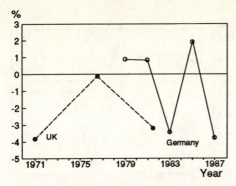
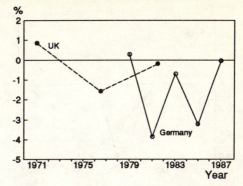

Figure 10.11. Contribution of changes in C to CO_2 emission: Germany and the UK.

Figure 10.12. Contribution of changes in $(I-A)^{-1}$ to CO_2 emission: Germany and the UK.

In Figure 10.11 we see that the ΔC elements are negative for the UK, while they become negative in Germany, though there is an odd 'blip' for the 1984-6 difference. This suggests that, overall, one might anticipate reductions in CO_2 emissions because of changing fuel mix and efficiency in the production part of the economy, *ceteris paribus*.

In Figure 10.12 the $\Delta(I-A)^{-1}$ elements are also mostly negative, also indicating that, *ceteris paribus*, over time interindustry trading is altering so as to reduce CO_2 emissions in both Germany and the UK.

Figure 10.13 shows that the Δu elements are mostly negative, with evidence for the UK that they have been falling. This suggests that the structure of final demand has been altering so that, *ceteris paribus*, CO_2 emissions have been falling in both countries. This almost certainly reflects the shift away from heavy industry and towards the provision of services.

In Figure 10.14 we see the impact of changing GDP on CO_2 emissions, *ceteris paribus*, though the ΔY component. Apart from the negative change in Germany for 1980-82, reflecting the world-wide economic recession, all of these components are positive, reflecting the generally maintained economic growth by both countries.

10 Input-Output Analysis of German and UK CO_2 Emissions

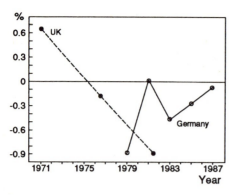

Figure 10.13. Contribution of changes in proportions of elements of **y** to CO_2 emission (**u**): Germany and the UK.

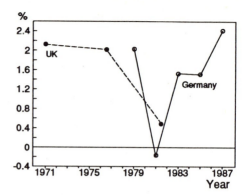

Figure 10.14. Contribution of changes in level of **y** to CO_2 emission (Y): Germany and the UK.

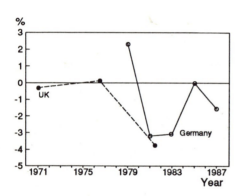

Figure 10.15. Contribution of total changes in production to CO_2 emission: Germany and the UK.

Figure 10.15 shows the total effect of changes in the production part of the economy on CO_2 emissions. Apart from the figure for Germany 1978-80, the overall effect has been mostly negative. That is, *ceteris paribus*, there has been a reduction in CO_2 emissions from the production part of the economy.

In Figure 10.16 we see the contribution to CO_2 emission changes by changing fuel prices, via the $\Delta \mathbf{P}$ term, with expenditure fixed. These indicate that overall, fuels have become less expensive in real terms over the past decade, causing increased contributions to CO_2 emissions, *ceteris paribus*.

Figure 10.17 shows the effect of the $\Delta \mathbf{Z}$ component on total CO_2 emissions. This represents the overall effect of the changing share of household consumer demand fuels in overall fuel final demand.

10.4.2 The Aggregate Decomposition   187

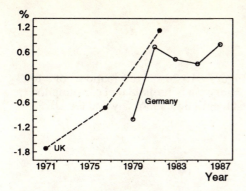

Figure 10.16. Contribution of changes in **P** to CO_2 emission: Germany and the UK.

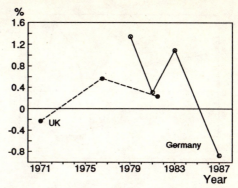

Figure 10.17. Contribution of changes in **Z** to CO_2 emission: Germany and the UK.

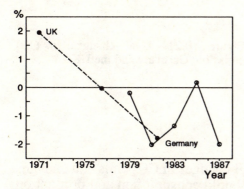

Figure 10.18. Contribution of changes in structure of household fuel purchases to CO_2 emission: Germany and the UK.

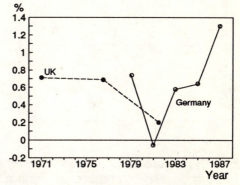

Figure 10.19. Contribution of changes in level of **y** to CO_2 emission: Germany and the UK.

In Figure 10.18 we see the effect of the changing structure of household fuel purchases on CO_2 emissions. We see that since 1977 this has been mostly negative in both countries, indicating that such this substitution between fuels by households has reduced CO_2 emissions below what they would have been otherwise.

Figure 10.19 shows the effect on CO_2 emission through household fuel use, resulting from the increasing purchasing power of consumers. This has been almost entirely positive.

In Figure 10.20 is displayed the overall effect on CO_2 emissions of the factors affecting household fuel use. The effects vary over time, but show no clear trend.

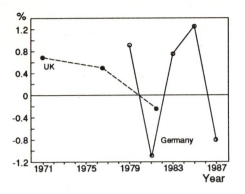

Figure 10.20. Contribution of total changes in household fuel consumption to CO_2 emission: Germany and the UK.

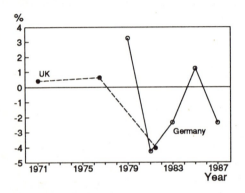

Figure 10.21. Total changes in CO_2 emission: Germany and the UK.

Finally, Figure 10.21 shows the overall changes in CO_2 emission changes in both countries, combining the household consumption and production uses of fuels. There is no clear evidence of an overall trend, reflecting the various and diverse influences discussed in the above decomposition.

10.5 Differences in CO_2 Emissions between Germany and the UK

Having detailed the structure of the changes in CO_2 emissions for Germany and the UK, over time, we now explore the differences between CO_2 emissions *between* Germany and the UK. In Tables 10.6 and 10.7 below, we examine the changes in CO_2 emissions for UK 1978 - Germany 1980, and UK 1984 - Germany 1984. Clearly, in the first comparison, the effects

10.5 Differences in CO_2 Emissions between Germany and the UK

of economic growth and technological change over one year for Germany will also be involved. In the second comparison, the differences should reflect the actual structural differences between the two economies.

From the 'Total' column of Table 10.6 we see that in 1980 Germany generated 17.75% more CO_2 than the UK in 1979. Of this, 4.56% is attributable to differences in fuel use for final demand directly, and 13.18% to the use of fuels in production. From Table 10.7 we see that by 1984, Germany was emitting 26.45% more CO_2 than the UK, attributable as 5.90% and 20.55% between final demand and production, respectively.

The great bulk of these differences can, in both cases, be attributed to larger GDP (ΔY). We see that these elements in the 'Total' row are all positive for both final demand consumption and production use of fuels, in both tables. This finding is precisely in line with intuition. Germany is a much larger economy, so one would expect that the contributions to CO_2 emission differences between the two countries would be positive when the levels of GDP were considered, *ceteris paribus*.

We also note that the elements ΔP and ΔC, in both tables, are mainly negative. This is an interesting finding, indicating that in most sectors the German economy is more efficient in its use of fuels than the UK economy. This probably reflects the newer capital stock in use in Germany, and also the higher standards of building construction required in that country.

We also note that the Δu terms are also mainly negative, for both consumption and production, and in both tables. Regarding consumption, this implies that the households mix of fuels in Germany is less CO_2 intensive than that in the UK. Concerning production, this finding indicates that the German mix of production generates less CO_2 than that of the UK, *ceteris paribus*.

The overall effect of the ΔL term is negative in Table 10.6, but positive in Table 10.7. This suggests that whereas the UK was initially more 'connected' through interindustry trading, by 1984 it was the German economy that was more 'connected'.

Final Demand	ΔP	ΔZ	Δu_C	ΔY_C	Sum	
6 Gas	-11.98	-1.92	-59.82	31.95	-40.77	
8 Coal extraction, coke ovens, etc	8.42	-96.28	11.59	33.89	-44.80	
12 Mineral oil processing	-48.01	67.27	0.44	48.39	65.48	
Total	-8.04	5.44	-4.29	11.90	4.56	
Production	ΔC	ΔL	Δu_P	ΔY_P	Sum	Total
1 Agriculture	32.82	-7.39	-22.26	52.05	53.84	53.84
2 Forestry & fishing	18.03	21.89	12.31	59.34	111.70	111.70
3 Electricity; fossil generation	0.00	0.00	0.00	0.00	0.00	0.00
4 Electricity; other generation	0.00	0.00	0.00	0.00	0.00	0.00
5 Electricity; distribution	-10.01	5.10	-53.05	38.67	-18.68	-18.68
6 Gas	0.27	0.93	-2.22	1.19	-0.14	-40.91
7 Water	-14.30	45.86	-132.70	36.09	-65.95	-65.95
8 Coal extraction, coke ovens, etc	47.13	7.39	4.18	12.24	72.97	28.18
9 Extraction of metalliferous ores	21.48	21.85	114.06	77.32	238.81	238.81
10 Extraction of mineral oil and gas	-64.04	539.28	-647.58	133.76	-71.48	-71.48
11 Chemical products	-25.14	3.66	7.17	46.57	32.05	32.05
12 Mineral oil processing	-13.03	1.66	0.05	5.15	-6.28	59.20
13 Processing of plastics	-23.71	-3.12	29.36	50.80	51.84	51.84
14 Rubber products	-10.77	-15.29	-6.43	43.02	10.36	10.36
15 Stone, clay, cement	-6.53	-4.36	5.73	49.41	42.85	42.85
16 Glass, ceramic goods	21.60	-2.61	2.31	53.62	74.69	74.69
17 Iron and steel, steel products	-141.99	40.92	146.33	71.67	80.91	80.91
18 Non-ferrous metals	3.56	80.32	-9.68	66.97	138.50	138.50
19 Foundries	-18.39	-7.34	-23.15	38.55	-9.59	-9.59
20 Production of steel etc	-39.98	0.64	-41.43	34.31	-44.12	-44.12
21 Mechanical engineering	-59.38	10.23	11.36	45.19	3.00	3.00
22 Office machines	-17.83	33.32	-16.38	52.25	49.20	49.20
23 Motor vehicles	-55.41	-16.49	66.39	53.46	38.93	38.93
24 Shipbuilding	-43.52	29.90	-54.48	40.26	-25.67	-25.67
25 Aerospace equipment	-28.28	2.62	-77.03	28.35	-69.23	-69.23
26 Electrical engineering	-29.76	-4.34	37.44	52.52	51.50	51.50
27 Instrument engineering	-25.28	-17.14	15.53	45.64	16.48	16.48
28 Engineers' small tools	-273.98	266.70	1304.34	311.66	1452.14	1452.14
29 Music instruments, toys, etc	-14.85	15.38	-59.32	39.06	-19.18	-19.18
30 Timber processing	44.32	33.01	-115.57	49.01	3.91	3.91
31 Wooden furniture	-15.30	12.46	118.83	72.63	187.09	187.09
32 Pulp, paper, board	-41.31	34.51	50.28	59.29	99.70	99.70
33 Paper and board products	-36.10	33.59	16.63	53.79	65.39	65.39
34 Printing and publishing	-16.53	29.09	-115.86	32.71	-69.09	-69.09
35 Leather, leather goods, footwear	-29.51	-15.69	-9.58	38.12	-16.35	-16.35
36 Textile goods	-27.14	-2.75	-22.29	38.67	-13.01	-13.01
37 Clothes	-34.34	-2.12	7.00	44.02	13.83	13.83
38 Food	8.51	-2.90	-8.13	49.70	46.36	46.36
39 Drink	-15.50	-10.46	-48.85	33.33	-39.86	-39.86
40 Tobacco	-34.99	-75.59	42.26	39.75	-31.75	-31.75
41 Construction	-0.87	13.36	7.86	56.55	72.98	72.98
42 Trade wholesale & retail	-4.94	-7.91	-34.59	40.15	-6.95	-6.95
43 Traffic & transport services	-2.69	1.80	-62.94	36.92	-26.65	-26.65
44 Telecommunications	-6.23	-22.08	-20.28	39.47	-8.74	-8.74
45 Banking, finance, insurance, etc	-41.31	24.86	50.36	57.67	88.25	88.25
46 Hotels, catering, etc	-5.18	16.81	-67.14	40.54	-14.95	-14.95
47 Other services	46.80	264.26	421.45	165.88	908.79	908.79
Total	-13.72	5.50	-10.91	33.80	13.18	17.75

Table 10.6. Attribution of difference in CO_2 emission between UK 1979 and Germany 1980 (%).

10.5 Differences in CO_2 Emissions between Germany and the UK

Final Demand	ΔP	ΔZ	Δu_C	ΔY_P	Sum	
6 Gas	-5.33	0.03	-62.36	31.77	-35.47	
8 Coal extraction, coke ovens, etc	-4.29	-89.89	31.95	33.72	-31.18	
12 Mineral oil processing	-78.51	84.32	12.32	50.74	62.44	
Total	-15.42	11.77	-3.97	14.78	5.90	
Production	ΔC	ΔL	Δu_P	ΔY_C	Sum	Total
1 Agriculture	66.23	1.79	-58.55	55.77	62.43	62.43
2 Forestry & fishing	111.51	1.09	7.67	72.83	194.40	194.40
3 Electricity; fossil generation	0.00	0.00	0.00	0.00	0.00	0.00
4 Electricity; other generation	0.00	0.00	0.00	0.00	0.00	0.00
5 Electricity; distribution	4.84	-5.50	-40.19	40.64	-0.17	-0.17
6 Gas	-0.76	0.68	-1.79	0.91	-1.07	-36.54
7 Water	-9.37	57.57	-142.71	36.66	-68.48	-68.48
8 Coal extraction, coke ovens, etc	23.05	-1.94	8.58	9.05	40.63	9.44
9 Extraction of metalliferous ores	-403.35	21.53	1856.55	414.25	1770.16	1770.16
10 Extraction of mineral oil and gas	10.86	266.26	-390.88	79.12	-95.07	-95.07
11 Chemical products	-8.29	12.76	2.55	50.48	58.47	58.47
12 Mineral oil processing	-8.71	-0.37	0.88	3.61	-4.66	57.79
13 Processing of plastics	-55.63	8.03	83.75	59.38	94.60	94.60
14 Rubber products	-43.64	1.38	11.36	43.68	13.09	13.09
15 Stone, clay, cement	13.16	-9.27	40.41	58.31	103.56	103.56
16 Glass, ceramic goods	55.06	-9.81	6.44	59.60	112.09	112.09
17 Iron and steel, steel products	-137.31	57.84	163.78	76.55	138.10	138.10
18 Non-ferrous metals	-14.48	31.72	-26.88	46.74	37.36	37.36
19 Foundries	-9.00	10.50	-6.57	47.30	43.63	43.63
20 Production of steel etc	-24.40	-12.59	-38.47	32.92	-40.85	-40.85
21 Mechanical engineering	-28.58	3.75	63.08	59.33	94.84	94.84
22 Office machines	-2.91	77.37	11.79	67.37	153.33	153.33
23 Motor vehicles	-51.78	2.37	129.05	70.41	142.52	142.52
24 Shipbuilding	-19.63	11.72	-61.75	37.15	-31.19	-31.19
25 Aerospace equipment	-11.65	0.22	-87.71	29.39	-67.65	-67.65
26 Electrical engineering	-13.50	-0.48	8.38	49.11	43.23	43.23
27 Instrument engineering	-14.81	1.47	34.18	53.89	75.51	75.51
28 Engineers' small tools	-114.71	73.49	681.75	185.61	793.97	793.97
29 Music instruments, toys, etc	-27.40	7.61	-13.36	42.19	9.89	9.89
30 Timber processing	5.37	9.44	-61.02	39.98	-7.81	-7.81
31 Wooden furniture	-40.16	15.39	57.22	56.87	89.83	89.83
32 Pulp, paper, board	-6.12	33.33	105.23	74.45	210.05	210.05
33 Paper and board products	-16.63	21.29	13.05	53.32	71.29	71.29
34 Printing and publishing	-2.85	16.49	-104.49	32.92	-60.50	-60.50
35 Leather, leather goods, footwear	2.23	-10.77	-17.86	44.66	18.35	18.35
36 Textile goods	22.11	25.64	-25.65	54.92	76.32	76.32
37 Clothes	14.78	33.31	17.56	61.55	129.14	129.14
38 Food	28.95	-7.08	-9.76	52.70	64.53	64.53
39 Drink	-0.61	-24.71	-36.96	36.34	-24.92	-24.92
40 Tobacco	5.68	-144.70	93.38	50.46	-2.33	-2.33
41 Construction	9.79	-13.42	-13.66	46.95	29.39	29.39
42 Trade wholesale & retail	16.50	-26.94	-42.20	38.12	-14.14	-14.14
43 Traffic & transport services	-15.34	-3.81	-43.58	36.45	-25.37	-25.37
44 Telecommunications	5.88	-73.48	4.14	36.76	-26.04	-26.04
45 Banking, finance, insurance, etc	-8.88	-38.51	44.02	49.99	47.21	47.21
46 Hotels, catering, etc	27.55	46.51	-99.66	49.83	17.63	17.63
47 Other services	59.17	-29.41	345.43	120.72	507.31	507.31
Total	-1.94	-1.35	-7.99	32.52	20.55	26.45

Table 10.7. Attribution of difference in CO_2 emission between UK 1984 and Germany 1984 (%).

10.6 Conclusions

The emission of CO_2 by the German and UK economies bears many great similarities. Both have an increasing proportion of emission attributable to direct final demand use of fossil fuels, and both have a falling proportion of CO_2 emissions attributable to direct and to indirect production of goods for final demand.

The proportion of CO_2 emission attributable to exports in both cases has moved to about 20%, with 80% of emissions attributable to domestic demand.

When imports are taken into account, some difference emerges between the two countries. For the UK the ratio of CO_2 responsibility to emissions has risen to 97%, while for Germany the ratio has been more stable, at about 93%. But both countries are, by this approximation, *responsible* for less CO_2 than they emit.

Regarding the decomposition of the changes in CO_2 emissions over time, we again see a reasonably close correspondence between Germany and the UK. For both there is evidence of improving fuel efficiency in production, and of decreasing connectedness through interindustry trading. In recent years there is also evidence that both countries have altered their mix of goods output towards lower overall CO_2 intensity.

Concerning final demand CO_2 emissions, there is evidence of *decreasing* efficiency in fuel use, but a shift in fuel purchasing patterns towards more gas, so a lower CO_2 intensity.

Comparing Germany and the UK, there is clear evidence that Germany is more efficient in its use of fuels than the UK, both by consumers and producers. Also, the German mix of output is less energy intensive than that of the UK.

A10 Appendix: Decomposition of Changes in CO_2 Emissions

A10.1 Introduction

In this Appendix are gathered the tables representing the decompositions of changes in CO_2 emissions by Germany and the UK, between successively dated input-output tables. These are as follows:

Germany	1978-80	**UK**	1968-74
	1980-82		1974-79
	1982-84		1979-84
	1984-86		
	1986-88		

All figures presented in the tables are annualised proportional rates of change.

The structure of each table is as for the tables presented in Chapter 10. That is, the first three rows represent the decomposition of the change in CO_2 emissions attributable directly to the final demand for fuels. These represent only the contributions by the three sectors:.Gas (6); Coal extraction, coke ovens, etc (8); Mineral oil processing (12). This is because fossil fuels are bought for final demand directly from these three sectors.

The fourth row, marked 'Total', shows the overall effect of changing domestic use of these fuels upon total household CO_2 emissions.

The succeeding 47 rows represent the deomposition of the change in CO_2 emissions attributable to production.

The final row shows the total components of changes in CO_2 emissions, summed over all sectors.

The interpretation of each column is as discussed in Section 10.4.

A10.2 Tables

There now follow the tables as described above.

A10 Appendix: Decomposition of Changes in CO_2 Emissions

Final Demand (% p.a.)	ΔP	ΔZ	Δu	ΔY	Sum	
6 Gas	-5.79	-0.73	1.97	2.56	-1.99	
8 Coal extraction, coke ovens, etc	-35.66	40.08	-6.93	1.68	-0.84	
12 Mineral oil processing	6.43	-4.61	0.66	2.64	5.12	
Total	-1.01	1.35	-0.18	0.75	0.91	

Production (% p.a.)	ΔC	ΔL	Δu	ΔY	Sum	Total
1 Agriculture	0.01	-0.51	-2.54	2.71	-0.34	-0.34
2 Forestry & fishing	-4.92	-0.32	7.27	2.78	4.82	4.82
3 Electricity; fossil generation	0.00	0.00	0.00	0.00	0.00	0.00
4 Electricity; other generation	0.00	0.00	0.00	0.00	0.00	0.00
5 Electricity; distribution	3.48	3.60	-2.25	2.81	7.64	7.64
6 Gas	0.18	-0.17	0.10	0.14	0.26	-1.74
7 Water	3.30	5.84	3.54	2.91	15.60	15.60
8 Coal extraction, coke ovens, etc	2.99	0.57	-7.03	1.70	-1.77	-2.60
9 Extraction of metalliferous ores	6.50	-0.76	28.34	3.24	37.32	37.32
10 Extraction of mineral oil and gas	3.36	3.25	38.96	3.41	48.98	48.98
11 Chemical products	1.42	1.02	-1.37	2.76	3.82	3.82
12 Mineral oil processing	-0.18	0.19	0.04	0.15	0.20	5.32
13 Processing of plastics	0.26	0.00	6.65	2.84	9.76	9.76
14 Rubber products	-0.51	-0.31	-2.67	2.70	-0.79	-0.79
15 Stone, clay, cement	0.27	0.40	-0.80	2.75	2.62	2.62
16 Glass, ceramic goods	4.01	-0.30	-0.02	2.80	6.49	6.49
17 Iron and steel, steel products	0.07	1.65	1.46	2.79	5.96	5.96
18 Non-ferrous metals	3.55	3.04	1.09	2.85	10.53	10.53
19 Foundries	1.44	2.01	10.15	2.93	16.52	16.52
20 Production of steel etc	0.66	4.83	2.62	2.86	10.97	10.97
21 Mechanical engineering	0.73	-0.26	-0.04	2.75	3.18	3.18
22 Office machines	-1.26	-3.62	8.47	2.81	6.40	6.40
23 Motor vehicles	0.68	-0.25	-2.69	2.72	0.45	0.45
24 Shipbuilding	-0.39	4.87	-3.35	2.77	3.89	3.89
25 Aerospace equipment	-1.80	-3.71	12.00	2.86	9.34	9.34
26 Electrical engineering	0.39	-1.38	1.69	2.76	3.46	3.46
27 Instrument engineering	0.54	-4.27	1.14	2.72	0.13	0.13
28 Engineers' small tools	0.85	0.47	-1.11	2.75	2.95	2.95
29 Music instruments, toys, etc	1.70	6.62	-12.59	2.72	-1.55	-1.55
30 Timber processing	-0.91	-0.37	10.51	2.88	12.11	12.11
31 Wooden furniture	0.61	0.07	-0.94	2.74	2.48	2.48
32 Pulp, paper, board	-0.65	1.36	4.46	2.82	7.99	7.99
33 Paper and board products	-0.59	4.22	1.51	2.82	7.96	7.96
34 Printing and publishing	0.37	-0.75	0.32	2.75	2.69	2.69
35 Leather, leather goods, footwear	1.53	-0.40	-1.05	2.75	2.83	2.83
36 Textile goods	1.26	0.12	-4.22	2.71	-0.13	-0.13
37 Clothes	1.40	-0.16	-3.90	2.71	0.05	0.05
38 Food	-0.31	-1.26	0.65	2.74	1.82	1.82
39 Drink	0.90	0.30	-2.03	2.74	1.90	1.90
40 Tobacco	-0.10	2.67	0.23	2.79	5.60	5.60
41 Construction	0.02	2.13	-0.30	2.77	4.62	4.62
42 Trade wholesale & retail	0.34	0.66	-1.86	2.74	1.88	1.88
43 Traffic & transport services	3.13	-7.18	-6.76	2.61	-8.20	-8.20
44 Telecommunications	-1.60	-6.88	5.42	2.73	-0.34	-0.34
45 Banking, finance, insurance, etc	-0.01	-1.04	0.94	2.75	2.64	2.64
46 Hotels, catering, etc	-0.84	-3.94	3.01	2.73	0.96	0.96
47 Other services	-2.26	-5.34	3.87	2.71	-1.02	-1.02
Total	0.88	0.31	-0.89	2.04	2.35	3.26

Table A10.1. Attribution of difference in CO_2 emission between Germany 1978 and Germany 1980.

A10.2 Tables

Final Demand (% p.a.)	ΔP	ΔZ	Δu	ΔY	Sum	
6 Gas	5.19	0.84	-3.77	-0.23	2.03	
8 Coal extraction, coke ovens, etc	-1.05	-1.46	-0.26	-0.09	-2.87	
12 Mineral oil processing	2.64	1.63	-8.91	-0.22	-4.86	
Total	0.72	0.30	-2.04	-0.06	-1.09	

Production (% p.a.)	ΔC	ΔL	Δu	ΔY	Sum	Total
1 Agriculture	-1.11	-5.01	9.79	-0.25	3.42	3.42
2 Forestry & fishing	-3.20	-1.45	-4.21	-0.23	-9.09	-9.09
3 Electricity; fossil generation	0.00	0.00	0.00	0.00	0.00	0.00
4 Electricity; other generation	0.00	0.00	0.00	0.00	0.00	0.00
5 Electricity; distribution	6.21	-8.39	2.16	-0.24	-0.27	-0.27
6 Gas	-0.19	-0.01	-0.19	-0.01	-0.41	1.63
7 Water	5.71	-5.45	-5.53	-0.24	-5.51	-5.51
8 Coal extraction, coke ovens, etc	-1.48	-0.36	-0.42	-0.14	-2.41	-5.28
9 Extraction of metalliferous ores	0.59	-3.38	-13.30	-0.22	-16.31	-16.31
10 Extraction of mineral oil and gas	3.95	-4.59	-25.01	-0.21	-25.86	-25.86
11 Chemical products	3.46	-5.08	0.05	-0.24	-1.81	-1.81
12 Mineral oil processing	-0.19	0.00	-0.46	-0.01	-0.66	-5.52
13 Processing of plastics	2.51	-8.07	3.37	-0.24	-2.42	-2.42
14 Rubber products	1.72	-3.97	-3.73	-0.23	-6.22	-6.22
15 Stone, clay, cement	2.75	-0.25	-8.97	-0.23	-6.71	-6.71
16 Glass, ceramic goods	-0.97	-1.52	-3.32	-0.23	-6.04	-6.04
17 Iron and steel, steel products	-2.82	-3.18	-6.19	-0.23	-12.41	-12.41
18 Non-ferrous metals	1.61	-11.20	3.43	-0.24	-6.40	-6.40
19 Foundries	-0.14	-5.67	-7.25	-0.23	-13.29	-13.29
20 Production of steel etc	-0.62	-6.88	-2.97	-0.23	-10.70	-10.70
21 Mechanical engineering	-0.70	-5.19	-1.29	-0.23	-7.41	-7.41
22 Office machines	-0.57	-6.01	9.44	-0.25	2.62	2.62
23 Motor vehicles	-0.37	-5.71	3.21	-0.24	-3.11	-3.11
24 Shipbuilding	-2.53	-7.81	10.32	-0.24	-0.27	-0.27
25 Aerospace equipment	-0.18	-5.02	7.54	-0.25	2.10	2.10
26 Electrical engineering	0.26	-8.15	-0.05	-0.23	-8.17	-8.17
27 Instrument engineering	1.42	-6.53	-4.78	-0.23	-10.11	-10.11
28 Engineers' small tools	-0.55	-5.42	-3.70	-0.23	-9.91	-9.91
29 Music instruments, toys, etc	0.73	-11.43	1.36	-0.23	-9.57	-9.57
30 Timber processing	-3.55	-4.64	-5.56	-0.22	-13.97	-13.97
31 Wooden furniture	0.89	-2.99	-7.39	-0.23	-9.73	-9.73
32 Pulp, paper, board	-1.81	-3.75	4.10	-0.24	-1.70	-1.70
33 Paper and board products	-0.84	-3.85	2.60	-0.24	-2.32	-2.32
34 Printing and publishing	0.44	-2.79	-0.11	-0.24	-2.70	-2.70
35 Leather, leather goods, footwear	-0.72	-4.05	-4.26	-0.23	-9.26	-9.26
36 Textile goods	0.65	-4.22	-2.01	-0.23	-5.81	-5.81
37 Clothes	1.10	-6.70	-5.72	-0.23	-11.55	-11.55
38 Food	-1.69	-4.28	0.59	-0.23	-5.61	-5.61
39 Drink	-3.48	-4.42	2.78	-0.24	-5.36	-5.36
40 Tobacco	0.53	-0.60	-6.54	-0.23	-6.84	-6.84
41 Construction	0.88	-4.50	-3.56	-0.23	-7.41	-7.41
42 Trade wholesale & retail	1.17	-3.43	-0.18	-0.24	-2.68	-2.68
43 Traffic & transport services	-2.37	-1.05	2.94	-0.24	-0.73	-0.73
44 Telecommunications	0.34	-8.97	6.88	-0.24	-1.99	-1.99
45 Banking, finance, insurance, etc	2.09	-5.93	3.42	-0.24	-0.66	-0.66
46 Hotels, catering, etc	0.38	-4.39	-0.36	-0.24	-4.60	-4.60
47 Other services	-2.20	-6.51	3.56	-0.24	-5.39	-5.39
Total	0.82	-3.85	0.02	-0.18	-3.19	-4.28

Table A10.2. Attribution of difference in CO_2 emission between Germany 1980 and Germany 1982.

A10 Appendix: Decomposition of Changes in CO_2 Emissions

Final Demand (% p.a.)	ΔP	ΔZ	Δu	ΔY	Sum	
6 Gas	0.79	0.48	3.95	2.12	7.34	
8 Coal extraction, coke ovens, etc	-0.74	11.06	-13.75	0.92	-2.50	
12 Mineral oil processing	2.00	2.57	-4.19	2.07	2.44	
Total	0.42	1.10	-1.36	0.58	0.75	

Production (% p.a.)	ΔC	ΔL	Δu	ΔY	Sum	Total
1 Agriculture	-1.57	0.58	0.54	2.15	1.69	1.69
2 Forestry & fishing	-2.30	0.98	2.57	2.17	3.42	3.42
3 Electricity; fossil generation	0.00	0.00	0.00	0.00	0.00	0.00
4 Electricity; other generation	0.00	0.00	0.00	0.00	0.00	0.00
5 Electricity; distribution	-6.55	-0.31	0.61	2.09	-4.14	-4.14
6 Gas	-0.13	-0.48	0.16	0.08	-0.38	6.97
7 Water	-6.24	2.72	-4.03	2.08	-5.47	-5.47
8 Coal extraction, coke ovens, etc	-1.74	-1.17	-16.22	1.08	-18.05	-20.55
9 Extraction of metalliferous ores	-8.07	-2.95	11.11	2.18	2.28	2.28
10 Extraction of mineral oil and gas	-5.59	5.28	-10.87	2.05	-9.13	-9.13
11 Chemical products	-5.23	-1.61	6.44	2.16	1.76	1.76
12 Mineral oil processing	-0.25	0.01	-0.18	0.09	-0.33	2.11
13 Processing of plastics	-5.51	-1.29	6.11	2.16	1.47	1.47
14 Rubber products	-5.73	-0.51	1.82	2.11	-2.31	-2.31
15 Stone, clay, cement	-8.70	-0.51	2.57	2.09	-4.55	-4.55
16 Glass, ceramic goods	-4.53	-0.56	2.71	2.13	-0.25	-0.25
17 Iron and steel, steel products	-4.21	-4.46	4.70	2.12	-1.85	-1.85
18 Non-ferrous metals	-4.53	-1.18	1.78	2.12	-1.82	-1.82
19 Foundries	-4.83	7.18	0.24	2.19	4.78	4.78
20 Production of steel etc	-2.25	-3.14	-8.84	2.01	-12.22	-12.22
21 Mechanical engineering	-3.65	-1.06	-3.70	2.07	-6.35	-6.35
22 Office machines	-7.68	-2.17	18.42	2.28	10.84	10.84
23 Motor vehicles	-4.56	-0.06	-0.74	2.10	-3.26	-3.26
24 Shipbuilding	-2.90	-1.85	-17.99	1.93	-20.82	-20.82
25 Aerospace equipment	-4.46	-5.93	-0.74	2.04	-9.08	-9.08
26 Electrical engineering	-4.71	-2.94	0.82	2.09	-4.75	-4.75
27 Instrument engineering	-4.89	1.19	0.23	2.12	-1.36	-1.36
28 Engineers' small tools	-4.46	-1.88	3.61	2.13	-0.60	-0.60
29 Music instruments, toys, etc	-4.30	4.83	-6.53	2.10	-3.90	-3.90
30 Timber processing	-7.03	-3.12	7.79	2.15	-0.22	-0.22
31 Wooden furniture	-4.44	1.15	-3.92	2.08	-5.13	-5.13
32 Pulp, paper, board	-5.16	-2.02	12.92	2.24	7.98	7.98
33 Paper and board products	-4.39	-2.80	5.00	2.14	-0.05	-0.05
34 Printing and publishing	-4.93	-2.26	4.24	2.13	-0.81	-0.81
35 Leather, leather goods, footwear	-4.79	3.03	-2.47	2.12	-2.11	-2.11
36 Textile goods	-4.42	0.44	-1.42	2.10	-3.30	-3.30
37 Clothes	-4.53	1.87	-1.73	2.11	-2.27	-2.27
38 Food	-2.13	-0.72	-3.20	2.09	-3.95	-3.95
39 Drink	-2.28	-0.05	-3.83	2.09	-4.06	-4.06
40 Tobacco	-4.56	-1.49	0.81	2.10	-3.13	-3.13
41 Construction	-5.47	-1.70	-0.79	2.07	-5.89	-5.89
42 Trade wholesale & retail	-3.91	-0.67	-0.54	2.10	-3.03	-3.03
43 Traffic & transport services	-2.69	1.71	-3.07	2.11	-1.94	-1.94
44 Telecommunications	-4.32	-1.86	0.37	2.10	-3.72	-3.72
45 Banking, finance, insurance, etc	-4.89	1.41	-0.68	2.11	-2.05	-2.05
46 Hotels, catering, etc	-3.38	0.93	-2.25	2.11	-2.59	-2.59
47 Other services	-4.91	-4.08	4.84	2.12	-2.03	-2.03
Total	-3.45	-0.69	-0.47	1.53	-3.08	-2.33

Table A10.3. Attribution of difference in CO_2 emission between Germany 1982 and Germany 1984.

Final Demand (% p.a.)	ΔP	ΔZ	Δu	ΔY	Sum	
6 Gas	0.44	-0.39	1.65	2.11	3.81	
8 Coal extraction, coke ovens, etc	-0.67	-3.59	0.37	1.06	-2.83	
12 Mineral oil processing	1.39	1.12	0.40	2.11	5.02	
	0.32	0.10	0.20	0.64	1.26	

Production (% p.a.)	ΔC	ΔL	Δu	ΔY	Sum	Total
1 Agriculture	3.57	-3.69	-0.45	2.16	1.60	1.60
2 Forestry & fishing	4.09	-3.16	3.99	2.22	7.14	7.14
3 Electricity; fossil generation	0.00	0.00	0.00	0.00	0.00	0.00
4 Electricity; other generation	0.00	0.00	0.00	0.00	0.00	0.00
5 Electricity; distribution	5.68	-7.81	-1.12	2.13	-1.11	-1.11
6 Gas	0.06	-0.23	0.05	0.07	-0.06	3.76
7 Water	5.56	-6.03	-0.19	2.16	1.49	1.49
8 Coal extraction, coke ovens, etc	2.51	0.17	0.38	1.09	4.15	1.32
9 Extraction of metalliferous ores	-11.56	-6.38	5.88	2.07	-9.99	-9.99
10 Extraction of mineral oil and gas	4.19	-12.60	-13.99	1.94	-20.47	-20.47
11 Chemical products	0.33	-5.55	-0.46	2.10	-3.58	-3.58
12 Mineral oil processing	-0.32	-0.04	0.01	0.07	-0.27	4.75
13 Processing of plastics	1.44	-5.68	3.28	2.16	1.19	1.19
14 Rubber products	2.91	-4.85	-0.35	2.14	-0.15	-0.15
15 Stone, clay, cement	2.95	-3.22	-1.41	2.14	0.47	0.47
16 Glass, ceramic goods	-0.01	-0.59	-0.28	2.15	1.27	1.27
17 Iron and steel, steel products	4.75	-2.81	-3.88	2.14	0.21	0.21
18 Non-ferrous metals	1.72	-9.12	1.68	2.11	-3.61	-3.61
19 Foundries	2.61	-12.04	5.59	2.14	-1.70	-1.70
20 Production of steel etc	2.29	-1.55	-1.98	2.15	0.91	0.91
21 Mechanical engineering	1.53	-2.92	2.62	2.18	3.41	3.41
22 Office machines	3.16	-7.78	10.46	2.24	8.09	8.09
23 Motor vehicles	1.79	-3.57	3.23	2.18	3.62	3.62
24 Shipbuilding	3.77	-1.53	-8.52	2.10	-4.17	-4.17
25 Aerospace equipment	3.14	-3.71	1.20	2.17	2.80	2.80
26 Electrical engineering	2.00	-4.88	3.88	2.18	3.18	3.18
27 Instrument engineering	1.80	-2.84	2.13	2.17	3.26	3.26
28 Engineers' small tools	2.70	-3.29	1.02	2.17	2.59	2.59
29 Music instruments, toys, etc	1.54	-5.93	-0.20	2.12	-2.47	-2.47
30 Timber processing	3.34	-2.50	1.63	2.19	4.66	4.66
31 Wooden furniture	2.97	-1.50	-5.59	2.12	-2.00	-2.00
32 Pulp, paper, board	1.11	-3.99	0.61	2.14	-0.14	-0.14
33 Paper and board products	-0.32	-4.36	0.36	2.12	-2.20	-2.20
34 Printing and publishing	2.15	-2.26	8.40	2.25	10.54	10.54
35 Leather, leather goods, footwear	3.55	-3.36	-4.00	2.12	-1.68	-1.68
36 Textile goods	-0.25	-3.24	-0.41	2.12	-1.78	-1.78
37 Clothes	1.39	-0.19	-4.69	2.13	-1.36	-1.36
38 Food	1.12	-3.45	0.31	2.14	0.12	0.12
39 Drink	5.53	-0.10	-5.93	2.16	1.65	1.65
40 Tobacco	2.95	-3.06	-3.32	2.13	-1.30	-1.30
41 Construction	1.86	-4.50	-4.14	2.09	-4.69	-4.69
42 Trade wholesale & retail	3.07	-5.48	1.29	2.15	1.04	1.04
43 Traffic & transport services	0.68	-0.72	-0.23	2.16	1.89	1.89
44 Telecommunications	4.02	-5.49	2.34	2.18	3.05	3.05
45 Banking, finance, insurance, etc	3.59	-3.57	0.64	2.17	2.83	2.83
46 Hotels, catering, etc	4.57	-4.38	-1.42	2.15	0.92	0.92
47 Other services	3.02	-2.60	-1.54	2.15	1.03	1.03
Total	1.98	-3.23	-0.27	1.51	0.00	1.26

Table A10.4. Attribution of difference in CO_2 emission between Germany 1984 and Germany 1986.

A10 Appendix: Decomposition of Changes in CO_2 Emissions

Final Demand (% p.a.)	ΔP	ΔZ	Δu	ΔY	Sum	
6 Gas	-10.74	0.04	7.41	4.20	0.91	
8 Coal extraction, coke ovens, etc	21.54	-14.23	-19.99	2.27	-10.42	
12 Mineral oil processing	3.19	-1.88	-7.68	4.11	-2.26	
Total	0.79	-0.89	-2.02	1.31	-0.81	

Production (% p.a.)	ΔC	ΔL	Δu	ΔY	Sum	Total
1 Agriculture	-4.59	-0.34	2.67	4.31	2.05	2.05
2 Forestry & fishing	-6.23	0.20	-1.35	4.21	-3.18	-3.18
3 Electricity; fossil generation	0.00	0.00	0.00	0.00	0.00	0.00
4 Electricity; other generation	0.00	0.00	0.00	0.00	0.00	0.00
5 Electricity; distribution	-5.70	-0.61	-2.60	4.17	-4.74	-4.74
6 Gas	-0.18	-0.48	0.19	0.11	-0.36	0.55
7 Water	-5.49	-3.45	-4.34	4.07	-9.21	-9.21
8 Coal extraction, coke ovens, etc	-1.87	-0.30	-17.14	1.95	-17.36	-27.78
9 Extraction of metalliferous ores	-13.06	2.89	3.72	4.26	-2.20	-2.20
10 Extraction of mineral oil and gas	-5.52	12.00	10.50	4.74	21.72	21.72
11 Chemical products	-1.83	-1.86	3.14	4.35	3.81	3.81
12 Mineral oil processing	-0.14	-0.01	-0.21	0.11	-0.25	-2.51
13 Processing of plastics	-4.76	0.53	3.74	4.36	3.87	3.87
14 Rubber products	-7.48	-1.89	2.28	4.22	-2.87	-2.87
15 Stone, clay, cement	-8.47	-0.42	1.23	4.21	-3.45	-3.45
16 Glass, ceramic goods	-12.88	0.75	1.12	4.15	-6.87	-6.87
17 Iron and steel, steel products	-1.74	-2.81	2.84	4.33	2.61	2.61
18 Non-ferrous metals	-6.27	4.06	-7.30	4.17	-5.33	-5.33
19 Foundries	-4.26	2.34	2.27	4.37	4.73	4.73
20 Production of steel etc	-6.18	-0.34	0.55	4.24	-1.73	-1.73
21 Mechanical engineering	-4.31	-1.64	-3.09	4.17	-4.88	-4.88
22 Office machines	-6.19	1.51	-4.87	4.16	-5.38	-5.38
23 Motor vehicles	-5.73	2.08	-0.74	4.27	-0.12	-0.12
24 Shipbuilding	-4.25	1.23	-10.70	4.08	-9.64	-9.64
25 Aerospace equipment	-4.97	-4.06	1.25	4.20	-3.59	-3.59
26 Electrical engineering	-6.27	-0.03	1.90	4.27	-0.12	-0.12
27 Instrument engineering	-5.96	-1.69	1.76	4.24	-1.64	-1.64
28 Engineers' small tools	-5.17	-0.93	1.90	4.28	0.08	0.08
29 Music instruments, toys, etc	-5.43	2.40	-0.75	4.29	0.50	0.50
30 Timber processing	-7.98	-0.52	6.01	4.33	1.83	1.83
31 Wooden furniture	-6.80	0.76	0.30	4.24	-1.50	-1.50
32 Pulp, paper, board	-7.29	1.59	7.22	4.41	5.92	5.92
33 Paper and board products	-5.00	0.04	1.56	4.29	0.89	0.89
34 Printing and publishing	-5.77	3.89	-1.82	4.29	0.59	0.59
35 Leather, leather goods, footwear	-5.13	3.11	-3.63	4.25	-1.40	-1.40
36 Textile goods	-4.42	-0.15	-1.08	4.24	-1.41	-1.41
37 Clothes	-5.13	1.45	-3.53	4.21	-3.01	-3.01
38 Food	-3.94	2.56	-3.20	4.27	-0.31	-0.31
39 Drink	-6.99	-0.05	-0.73	4.20	-3.56	-3.56
40 Tobacco	-6.07	0.90	-3.77	4.18	-4.77	-4.77
41 Construction	-6.79	-0.94	-0.39	4.19	-3.93	-3.93
42 Trade wholesale & retail	-5.29	-0.54	-0.42	4.23	-2.02	-2.02
43 Traffic & transport services	-12.85	6.42	6.93	4.40	4.90	4.90
44 Telecommunications	-5.50	3.55	1.46	4.37	3.87	3.87
45 Banking, finance, insurance, etc	-6.26	0.28	-0.69	4.22	-2.45	-2.45
46 Hotels, catering, etc	-5.95	-0.31	0.09	4.23	-1.94	-1.94
47 Other services	-6.14	-2.53	2.76	4.24	-1.66	-1.66
Total	-3.80	-0.04	-0.65	2.93	-1.56	-2.37

Table A10.5. Attribution of difference in CO_2 emission between Germany 1986 and Germany 1988.

Final Demand (% p.a.)	ΔP	ΔZ	Δu	ΔY	Sum	
6 Gas	11.76	0.06	1.39	0.95	14.16	
8 Coal extraction, coke ovens, etc	0.48	0.03	-9.62	2.51	-6.61	
12 Mineral oil processing	-18.77	-1.77	20.87	3.12	3.45	
Total	-1.72	-0.23	1.94	0.71	0.69	
Production (% p.a.)	ΔC	ΔL	Δu	ΔY	Sum	Total
1 Agriculture	-4.95	2.60	-12.00	2.70	-11.64	-11.64
2 Forestry & fishing	8.45	5.43	-11.90	3.14	5.12	5.12
3 Electricity; fossil generation	0.00	0.00	0.00	0.00	0.00	0.00
4 Electricity; other generation	0.00	0.00	0.00	0.00	0.00	0.00
5 Electricity; distribution	-2.73	0.68	0.67	2.70	1.32	1.32
6 Gas	-17.36	-1.28	2.56	1.76	-14.33	-0.17
7 Water	-5.18	0.66	10.32	2.91	8.72	8.72
8 Coal extraction, coke ovens, etc	-0.21	0.12	-0.65	0.17	-0.58	-7.18
9 Extraction of metalliferous ores	-9.19	10.46	0.53	2.83	4.63	4.63
10 Extraction of mineral oil and gas	0.00	0.00	0.00	0.00	0.00	0.00
11 Chemical products	-5.00	3.78	4.78	2.83	6.40	6.40
12 Mineral oil processing	-1.71	-0.35	3.13	0.47	1.54	4.99
13 Processing of plastics	-5.81	1.73	21.59	3.30	20.81	20.81
14 Rubber products	-9.02	2.53	1.20	2.73	-2.55	-2.55
15 Stone, clay, cement	-5.63	0.66	12.37	2.97	10.36	10.36
16 Glass, ceramic goods	-10.16	0.91	0.10	2.70	-6.44	-6.44
17 Iron and steel, steel products	-0.50	1.18	-2.01	2.68	1.34	1.34
18 Non-ferrous metals	-3.03	3.56	5.52	2.81	8.86	8.86
19 Foundries	-7.16	-5.32	73.12	6.31	66.96	66.96
20 Production of steel etc	-1.80	2.84	-9.13	2.71	-5.38	-5.38
21 Mechanical engineering	-3.13	2.50	1.00	2.74	3.11	3.11
22 Office machines	-6.65	-6.84	23.19	3.43	13.14	13.14
23 Motor vehicles	-0.45	-6.21	-0.63	2.65	-4.64	-4.64
24 Shipbuilding	-2.60	-2.21	-2.97	2.63	-5.15	-5.15
25 Aerospace equipment	-5.68	-5.60	5.38	2.76	-3.14	-3.14
26 Electrical engineering	-3.93	-1.31	0.51	2.68	-2.05	-2.05
27 Instrument engineering	-3.54	-0.23	-0.07	2.68	-1.15	-1.15
28 Engineers' small tools	-3.06	0.22	14.03	3.01	14.20	14.20
29 Music instruments, toys, etc	-4.94	1.20	-0.78	2.68	-1.84	-1.84
30 Timber processing	-2.12	10.52	3.11	2.95	14.46	14.46
31 Wooden furniture	-1.71	4.86	-4.82	2.80	1.13	1.13
32 Pulp, paper, board	-5.71	1.68	5.51	2.80	4.28	4.28
33 Paper and board products	-3.49	2.38	0.99	2.74	2.62	2.62
34 Printing and publishing	-4.87	0.34	-2.20	2.65	-4.07	-4.07
35 Leather, leather goods, footwear	-2.07	0.93	-4.42	2.69	-2.88	-2.88
36 Textile goods	-2.08	1.49	-1.76	2.72	0.37	0.37
37 Clothes	-3.17	0.41	-0.36	2.71	-0.41	-0.41
38 Food	-4.03	1.40	1.54	2.73	1.64	1.64
39 Drink	-6.81	0.55	6.51	2.81	3.06	3.06
40 Tobacco	-4.68	-1.53	2.25	2.69	-1.28	-1.28
41 Construction	-5.01	6.09	-6.73	2.78	-2.87	-2.87
42 Trade wholesale & retail	-3.84	1.63	1.15	2.74	1.68	1.68
43 Traffic & transport services	3.09	-0.26	2.55	2.80	8.17	8.17
44 Telecommunications	-8.25	0.67	2.02	2.77	-2.79	-2.79
45 Banking, finance, insurance, etc	-9.77	1.64	0.18	2.70	-5.25	-5.25
46 Hotels, catering, etc	-9.87	7.52	0.18	2.76	0.60	0.60
47 Other services	-10.07	1.96	0.18	2.70	-5.24	-5.24
Total	-3.83	0.84	0.65	2.12	-0.30	0.39

Table A10.6. Attribution of difference in CO_2 emission between UK 1968 and UK 1974.

A10 Appendix: Decomposition of Changes in CO_2 Emissions

Final Demand (% p.a.)	ΔP	ΔZ	Δu	ΔY	Sum	
6 Gas	0.14	-0.04	6.24	2.56	8.90	
8 Coal extraction, coke ovens, etc	3.76	1.55	-13.17	2.61	-5.26	
12 Mineral oil processing	-6.37	3.09	2.40	2.32	1.44	
Total	-0.73	0.57	-0.03	0.69	0.50	

Production (% p.a.)	ΔC	ΔL	Δu	ΔY	Sum	Total
1 Agriculture	0.12	-2.83	-5.55	2.63	-5.63	-5.63
2 Forestry & fishing	-15.19	-4.44	2.94	2.75	-13.95	-13.95
3 Electricity; fossil generation	0.00	0.00	0.00	0.00	0.00	0.00
4 Electricity; other generation	0.00	0.00	0.00	0.00	0.00	0.00
5 Electricity; distribution	0.39	-0.37	1.31	2.73	4.06	4.06
6 Gas	-1.80	-0.51	0.59	0.24	-1.47	7.42
7 Water	1.77	24.49	-33.12	3.64	-3.21	-3.21
8 Coal extraction, coke ovens, etc	-0.11	0.14	-0.72	0.14	-0.55	-5.81
9 Extraction of metalliferous ores	3.37	-3.17	-2.98	2.69	-0.09	-0.09
10 Extraction of mineral oil and gas	8.60	-188.1	261.2	17.71	99.44	99.44
11 Chemical products	-0.80	-2.75	-0.60	2.67	-1.49	-1.49
12 Mineral oil processing	-0.59	-0.38	0.44	0.42	-0.11	1.33
13 Processing of plastics	0.22	-3.08	8.77	2.84	8.75	8.75
14 Rubber products	1.46	-0.43	-1.08	2.71	2.66	2.66
15 Stone, clay, cement	-1.98	-0.08	-5.26	2.64	-4.69	-4.69
16 Glass, ceramic goods	-1.14	-0.57	-4.10	2.66	-3.15	-3.15
17 Iron and steel, steel products	0.60	0.15	0.42	2.72	3.89	3.89
18 Non-ferrous metals	-1.96	-1.71	-3.03	2.64	-4.05	-4.05
19 Foundries	3.00	-0.05	-22.69	2.80	-16.94	-16.94
20 Production of steel etc	-0.97	-3.41	3.54	2.73	1.89	1.89
21 Mechanical engineering	4.00	-5.36	0.57	2.70	1.91	1.91
22 Office machines	-0.24	-3.65	-1.94	2.66	-3.18	-3.18
23 Motor vehicles	-2.95	0.33	0.10	2.67	0.15	0.15
24 Shipbuilding	0.04	-2.95	-4.17	2.64	-4.45	-4.45
25 Aerospace equipment	-0.22	-0.56	-3.53	2.66	-1.66	-1.66
26 Electrical engineering	-0.93	-6.79	-0.98	2.64	-6.07	-6.07
27 Instrument engineering	-1.37	-3.91	1.55	2.68	-1.05	-1.05
28 Engineers' small tools	-0.12	-5.08	0.77	2.68	-1.75	-1.75
29 Music instruments, toys, etc	-3.11	-9.48	4.57	2.75	-5.27	-5.27
30 Timber processing	-1.93	-2.12	-7.26	2.59	-8.72	-8.72
31 Wooden furniture	-2.76	-4.30	9.23	2.82	4.99	4.99
32 Pulp, paper, board	-1.13	0.90	-6.29	2.66	-3.86	-3.86
33 Paper and board products	-1.76	-2.89	-1.45	2.64	-3.46	-3.46
34 Printing and publishing	0.20	-3.98	-0.29	2.67	-1.41	-1.41
35 Leather, leather goods, footwear	-2.15	-5.47	-2.12	2.62	-7.12	-7.12
36 Textile goods	-1.23	-3.90	-4.54	2.61	-7.07	-7.07
37 Clothes	0.56	-6.21	-0.08	2.66	-3.07	-3.07
38 Food	-1.54	0.11	-2.26	2.68	-1.01	-1.01
39 Drink	-1.49	-2.10	3.36	2.72	2.49	2.49
40 Tobacco	0.19	-0.45	-0.48	2.70	1.96	1.96
41 Construction	-0.96	-6.32	0.17	2.66	-4.45	-4.45
42 Trade wholesale & retail	2.14	-2.76	1.41	2.73	3.52	3.52
43 Traffic & transport services	2.72	0.87	-5.96	2.72	0.35	0.35
44 Telecommunications	1.07	-2.13	0.76	2.71	2.42	2.42
45 Banking, finance, insurance, etc	0.55	-9.01	3.81	2.72	-1.92	-1.92
46 Hotels, catering, etc	1.59	-0.47	-1.45	2.72	2.38	2.38
47 Other services	-2.83	-16.78	-2.81	2.64	-19.78	-19.78
Total	-0.13	-1.59	-0.18	2.02	0.13	0.63

Table A10.7. Attribution of difference in CO_2 emission between UK 1974 and UK 1979.

Final Demand (% p.a.)	ΔP	ΔZ	Δu	ΔY	Sum	
6 Gas	0.80	0.06	1.07	0.68	2.61	
8 Coal extraction, coke ovens, etc	1.13	7.26	-17.23	0.69	-8.15	
12 Mineral oil processing	5.99	-0.70	-6.57	0.62	-0.67	
Total	1.11	0.23	-1.78	0.20	-0.24	

Production (% p.a.)	ΔC	ΔL	Δu	ΔY	Sum	Total
1 Agriculture	-5.50	-3.81	9.46	0.73	0.88	0.88
2 Forestry & fishing	-12.53	3.16	0.04	0.68	-8.65	-8.65
3 Electricity; fossil generation	0.00	0.00	0.00	0.00	0.00	0.00
4 Electricity; other generation	0.00	0.00	0.00	0.00	0.00	0.00
5 Electricity; distribution	-4.21	-0.27	-1.95	0.68	-5.75	-5.75
6 Gas	0.04	-0.06	0.03	0.02	0.03	2.64
7 Water	-1.10	-3.40	0.91	0.68	-2.91	-2.91
8 Coal extraction, coke ovens, etc	0.39	1.30	-1.48	0.06	0.27	-7.88
9 Extraction of metalliferous ores	7.27	0.86	-42.13	0.88	-33.13	-33.13
10 Extraction of mineral oil and gas	-3.83	13.16	11.56	0.80	21.69	21.69
11 Chemical products	-4.34	-3.33	3.36	0.69	-3.62	-3.62
12 Mineral oil processing	-0.50	0.37	-0.95	0.09	-0.99	-1.65
13 Processing of plastics	1.14	-3.88	-3.12	0.67	-5.19	-5.19
14 Rubber products	3.17	-3.76	-3.96	0.68	-3.88	-3.88
15 Stone, clay, cement	-4.75	-0.08	-6.92	0.66	-11.09	-11.09
16 Glass, ceramic goods	-6.20	0.05	-0.79	0.68	-6.26	-6.26
17 Iron and steel, steel products	-7.61	-2.09	-1.87	0.67	-10.90	-10.90
18 Non-ferrous metals	-0.04	1.46	5.88	0.73	8.03	8.03
19 Foundries	-5.80	-0.87	-6.29	0.66	-12.31	-12.31
20 Production of steel etc	-5.10	-0.90	-4.97	0.66	-10.31	-10.31
21 Mechanical engineering	-8.70	-0.82	-8.05	0.65	-16.92	-16.92
22 Office machines	-4.93	-7.99	6.84	0.71	-5.37	-5.37
23 Motor vehicles	-4.76	-4.59	-4.08	0.65	-12.78	-12.78
24 Shipbuilding	-6.62	-0.56	-1.09	0.68	-7.59	-7.59
25 Aerospace equipment	-7.53	-3.38	6.30	0.71	-3.90	-3.90
26 Electrical engineering	-4.68	-4.77	4.62	0.69	-4.15	-4.15
27 Instrument engineering	-3.74	-5.00	-4.06	0.65	-12.15	-12.15
28 Engineers' small tools	-4.57	0.92	9.85	0.74	6.94	6.94
29 Music instruments, toys, etc	0.07	-0.55	-11.15	0.67	-10.97	-10.97
30 Timber processing	2.85	-0.41	-6.82	0.70	-3.68	-3.68
31 Wooden furniture	2.24	-1.35	0.53	0.70	2.12	2.12
32 Pulp, paper, board	-8.35	-0.26	1.73	0.69	-6.19	-6.19
33 Paper and board products	-4.86	-1.00	3.53	0.69	-1.64	-1.64
34 Printing and publishing	-5.03	1.17	-2.96	0.68	-6.14	-6.14
35 Leather, leather goods, footwear	-8.17	-2.00	-1.49	0.66	-11.00	-11.00
36 Textile goods	-9.99	-4.90	-2.11	0.66	-16.35	-16.35
37 Clothes	-8.99	-5.69	-3.83	0.64	-17.86	-17.86
38 Food	-4.60	-1.28	-0.85	0.67	-6.06	-6.06
39 Drink	-5.73	1.04	-3.96	0.67	-7.99	-7.99
40 Tobacco	-7.70	5.29	-9.11	0.68	-10.84	-10.84
41 Construction	-3.67	1.67	1.65	0.70	0.36	0.36
42 Trade wholesale & retail	-5.66	2.46	1.86	0.70	-0.64	-0.64
43 Traffic & transport services	0.74	1.47	-4.31	0.69	-1.41	-1.41
44 Telecommunications	-5.20	8.49	-2.16	0.72	1.86	1.86
45 Banking, finance, insurance, etc	-5.44	7.75	0.98	0.72	4.00	4.00
46 Hotels, catering, etc	-6.98	-4.96	2.27	0.68	-8.99	-8.99
47 Other services	-4.99	12.77	-0.88	0.75	7.65	7.65
Total	-3.19	-0.19	-0.89	0.49	-3.77	-4.02

Table A10.8. Attribution of difference in CO_2 emission between UK 1979 and UK 1984.

Part V

Scenarios

11 Scenario Simulations

11.1 Introduction

In this chapter we shall formulate various scenarios for economic restructuring, for Germany and the UK, and assess the associated CO_2 emissions. The method we follow is an application of our basic input-output equation for CO_2 emissions, derived in Section 8.2.4; viz:

$$C = [e'P'Z + (e'C' + m')(I - A)^{-1}]y. \qquad (8.16)$$

In particular we shall explore the effects of altering P, C and y. In detail, the simulation experiments described in this chapter reflect possible CO_2 control strategies and concentrate on the following issues:

1. *Final Demand*. We explore the effects of altering the structure of final demand, for a given growth rate of total GDP (Section 11.2).

2. *Energy Efficiency*. We change the energy efficiency by changing the technologies used for specific purposes, by going from average to best practice (Section 11.3).

3. *Inter-Fuel Substitution*. This uses switching to a mix of fuel inputs with lower CO_2 emissions. Possibilities include moving from fossil fuels to nuclear, wind, solar, or hydropower generated electricity (Section 11.4).

4. *Trend Forecasts*. These are based on some highly tentative trend extrapolation exercises (Section 11.5).

5. *'Plausible' Scenarios*. These summarize our most important simulation results (Section 11.6). They also provide insights with respect to the Toronto targets, and forms the platform for a new minimum disruption approach to scenario analysis, to be presented in Chapter 12. (The base year of all scenarios in this chapter is 1984, for both Germany and the UK.)

Finally, in Section 11.7 we draw our conclusions.

11.2 Changing the Structure of Final Demand

This section presents scenarios that assess how a change in the mix of final demand (**u**), given the growth rate of GDP (Y), influences CO_2 emissions. We assume that what will happen to final demand is determined exogenously (e.g. over a given period), and then we use the model to forecast the consequent levels of total output (**x**), employment and CO_2 emission. In this section, we stipulate a constant technology, including constant energy efficiency.

11.2.1 Final Demand for Non-Fuel Goods

Our model assumptions imply that a one percent change in the level of total final demand (i.e. GDP), given the structure of final demand, reduces CO_2 emissions by one percent. The percentage impact on CO_2 emissions of a one percent change of a component of final demand, given the level of all other components of final demand, is given by the respective emission elasticities (see Section 9.5.6).

Therefore it remains to check the impact on emissions of a change in the mix of final demand, given an exogenous growth rate of total final demand.

We first analyze the possibilities for CO_2 reduction by altering the structure of final demand for non-fuel goods, for a given growth rate of GDP.

We start with a strong abatement scenario and assume a significant expansion of the 'tertiary sectors' (i.e. services).

Previous studies typically assume a GDP growth rate of around 2% per annum over the next century (see Chapter 3 above for more detail). We follow this assumption and posit a GDP growth rate of 2% p.a., over 20 years.

On top of that, we posit additionally an 8% p.a. increase in the demand for commodities produced in sectors 44 to 47 (i.e. the tertiary sectors: telecommunications, banking, finance, insurance etc., hotels, catering etc. and other services). The final demand for commodities produced in sectors 5 to 12 (i.e. electricity distribution, gas, water, coal extraction, coke ovens etc., extraction of metalliferous ores, extraction of mineral oil and gas, chemical products and mineral oil processing) is assumed to be cut proportionally such that the basic growth rate of GDP (i.e. 2%) is maintained.

We also calculate the effects on employment. We do this by calculating the effect of the changed final demand (**y**) on total output (**x**), using the Leontief inverse. We then calculate the new total employment, using the known labour/output ratios for each sector, and for both countries.

Table 11.1. presents our simulation results.

% per annum	Germany	UK
Change in CO_2 emissions	-10.22	-3.45
Change in employment	3.14	4.60

Table 11.1. Change in CO_2 emissions with tertiary sectors expansion, given GDP growth rate of 2% p.a.

This radical change from the present structure to a post-industrial energy saving society yields impressive reductions in CO_2 emissions, far beyond the Toronto targets, and in spite of a significant growth in the level of total GDP. Notice that the changing structure of final demand corresponds to an increase in employment of more than 3% p.a. in both countries. (Of course, this assumes that the output/labour ratio for each sector is unchanged. In reality, these ratios have been falling.)

Whether economic agents would be willing to change their behaviour as required in this example is of course an open question. We will therefore assess other ways to curtail emissions.

11.2.2 Final Demand for Fuels

We now assess the possibilities for CO_2 reduction by altering the structure of final demand for fuels, for a given level of GDP.

First, we assume a 10% p.a. increase of the final demand for gas, but a corresponding decrease in the final demand for coal, such that again the level of GDP remains constant.

Next, we look at the impact of a 10% p.a. increase of the final demand for gas, and a corresponding decrease in the final demand for oil, such that the level of GDP remains constant.

The results in Table 11.2 show that the impact of a 10% increase of the final demand for gas, accompanied by a corresponding decrease in oil or coal consumption, is small. In Germany, diminished coal use in favour of gas decreases CO_2 emissions by only 0.04%, in the UK by 0.24%. Substitution of oil by gas in neither country significantly changes CO_2 emissions. Employment effects are negligible.

% per annum	Germany	UK
Change in CO_2 emissions	-0.04	-0.25
Change in employment	-0.00	-0.02

(a) Substitution of coal by gas: A 10% increase of final demand for gas and decrease for coal, such that the level of GDP remains constant.

% per annum	Germany	UK
Change in CO_2 emissions	-0.04	-0.25
Change in employment	-0.00	-0.15

(b) Substitution of oil by gas. A 10% increase of final demand for gas and decrease for oil such that the level of GDP remains constant.

Table 11.2. Possibilities for CO_2 emission reduction by altering the structure of final demand for fuels, for a given level of GDP.

The results presented above suggest that only small CO_2 savings can be won by inter-fossil fuel substitution by households, given the level of GDP. So we might presume that it is not a switch in the final demand for fossil fuels, but a reduction of fuel use (or a turn to nuclear or renewables) that might eventually help to decrease CO_2 emissions.

11.2.3 From Private to Public Transport

This section presents some simulation experiments modelling the impact of changes in final demand, with a partial switch from private transport to mass transit.

In recent years, in Germany and in the UK, road transport has gained market share at the expense of rail transport. Although attitudes to private transportation appear to be changing in some quarters, the demand for private transportation is still strongly determined by the level of GDP.

However, new technologies of mass transit, such as high speed trains and modern urban trams, can increase the desirability and speed of public transport to and from work, and between cities. Increased use of telecommunications and computer networks can also reduce the need for physical movements.

In a previous study of this topic, Victor [1972:200-209] examines the implications of a 50% cut in private car use in favour of public transport. He estimates a new final demand vector, with changes in petroleum products, vehicles and parts, rubber products, distributive services and transportation. His results suggest a decline in ecological costs, with the change in the final demand for petroleum products accounting for most of this. Guentenspenger [1990] presents evidence on how the vintage structure of the stock of passenger cars determines the level of CO_2 emissions. Finally, a report presented by Energetics Inc. in 1989, and quoted by Streb [1989], suggests that primary fossil energy demand in the transportation sector (i.e. oil) may be more than halved during the next 60 years.

11.2.4 Simulation Results

We now turn to our own scenarios. In our simulation exercise we assume that private car use in both the UK and Germany is curtailed by 5 percent per annum, and is replaced by public transport.

More specifically we stipulate the following exogenous changes in final demand:

Sector 12: Mineral Oil Processing	-3.0 %
Sector 14: Rubber Products	-0.5 %
Sector 23: Motor Vehicles	-5.0 %
Sector 41: Construction	-1.0 %
Sector 43: Traffic and Transport Services	-5.0 %
Sector 44: Telecommunications	0.5 %

Because of these changes in final demand total CO_2 emissions are cut back by roughly one percent per annum in both countries (see Table 11.3). Employment decreases slightly.

% per annum	Germany	UK
Change in CO_2 emissions	-1.07	-0.94
Change in employment	-0.57	-0.35

Table 11.3. CO_2 emission reductions due to altering the structure of final demand: private to public transport.

In the long term, it may be desirable to disaggregate the traffic and transport sector in our model, to link transportation costs to investments in infrastructure, and to account for maintenance costs as depending on the volume of traffic.

11.2.5 Conclusion

This section has shown that a marked shift away from the primary sectors of the economy, towards a large and rapidly expanding services sector, and a dramatic switch from private to public transport, can cut CO_2 levels to reach the Toronto targets, even with present levels of energy efficiency. We do not, however, expect these drastic changes to take place, because of the heavy adjustment costs, and habit persistence of the economic agents.

Therefore, the next section assesses the impact of changing energy efficiency, given the structure and level of output.

11.3 Changing the Energy Efficiency

If CO_2 emissions are to be cut, more efficient fossil fuel use can play an important role, because this reduces CO_2 emissions without reducing economic growth.

There is a broad strand of technical literature dealing with energy technologies for reducing emissions of greenhouse gases (see OECD [1989a] for a survey). Comprehensive studies by Maier and Angerer [1986] and by Kolb et al. [1989] present many technical details for the rational use of energy in Germany, but they do not assess the impact of behavioural changes of the economic agents, nor of any changes in the demand for energy.

Economic studies surveyed by Hoeller et al. [1991] typically assume a baseline growth in autonomous energy efficiency, between 0-2% p.a. over the next century, usually with stronger growth in the next few decades. For example, Reilly et al. [1987] assume long-term average growth rates of energy efficiency of 0.8-1.0% p.a.

In our simulation experiments, we analyze the potential for abating CO_2 emissions due to increased fuel efficiency; i.e. getting the same output for less fuel input. Better fuel efficiency may be due to changes in engineering and maintenance practices, or to the installation of more efficient capital as plant and equipment. In the latter case, fuel can be used more efficiently by, e.g., capital retrofitting (see Blitzer et al. [1990a, 1990b]).

We disregard the latter effects and concentrate on autonomous changes in the efficiency of industrial fuel use (i.e. the **C** matrix, see Section 11.3.1) and the efficiency of direct final demand fuel use (i.e. the **P** matrix, see Section 11.3.2).

11.3.1 Efficiency of Industrial Fuel Use

Advanced technologies can curtail fossil fuel consumption and CO_2 emissions compared with conventional technologies. The scenarios described later in this section analyze a turn from conventional average practice technologies, to best practice technologies. However, before embarking on simulation results, we present a set of examples which show that current best practice technologies can yield lower emissions than the average practice technologies now in use.

11.3.1.1. Electricity Generation

There are various advanced electricity generation technologies that offer a considerable scope for improvement in the efficiency of energy generation processes.

Improvements in energy utilization may be due to coal-fired combined cycles, with pressurized fluidised bed combustion, due to coal-fired integrated gasification combined cycles, or due to other options under development. These advanced technologies are cost-effective in the long term, and they exhibit high thermal efficiency and fuel flexibility.

If there is sufficient demand for heat or steam at an industrial, commercial or residential site, then central power generation may be decentralized, by turning to gas-fired combined heat and power generation, that operates at 85% efficiency [Graham-Bryce 1989].

A comparison of power station efficiencies, for different types of fossil-fuel fired power plants, shows a remarkable range of efficiency. A conventional steam turbine power plant with pulverized-coal firing, desulphurisation and NO_x removal, has 30-35% efficiency, while a modern combined cycle power plant, with natural-gas firing, has an efficiency of more than 50%.

As an example, we look at natural gas combined cycle power generation. In a simulation exercise we find that the replacement of conventional coal fired power plants by coal fired combined cycle plants would reduce CO_2 emissions in both countries by about 20 M tonnes (i.e. 19.8 M tonnes in Germany (3.0%) and 18.8 M tonnes in the UK (3.6%)). (The results for Germany are in line with Wagner and Kolb [1989].)

Higher reductions in CO_2 emissions would be available if all existing coal fired power plants were replaced by natural gas combined cycle fired power plants. From a technical perspective, the technological upper limit exceeds 100 M tonnes CO_2 in each country (Germany: 15%; UK: 19%).

Summing up, in electricity generation a switch from average technologies to advanced generation technologies, such as combined heat and power generation, or natural gas combined cycle power generation, is near commercialisation and can significantly decrease CO_2 emissions.

11.3.1.2 Iron and Steel Industry

The iron and steel industry is one of the most energy-intensive sectors in the economy. As combustible forms of secondary energy resources (blast-furnace gas, coke-oven gas, converter gases) are produced in this sector, we expect considerable future fuel savings per unit of output.

Energy-saving options include integration processes, dry quenching of coke, waste heat recovery measures, introduction of blast furnace top gas pressure turbines, recovery of converter gas, and switching from oil to coke and natural gas in blast furnaces (see OECD [1988] for a survey).

New options are being developed currently. For example, a new regenerative burner system for all types of furnaces, offered by Stordy Combustion Engineering Ltd. in Wolverhampton, UK, allows fuels savings of 35-40% in steel coil annealing, of 50-60% in stainless steel strip heat treatment, and of 45-55% in aluminium melting.

Summing up, fuel savings of 30-60% can typically be obtained.

In France, highly efficient electrical furnaces are being developed by Stinn Codara in Patin. They save half of the primary energy needed for metal working furnaces, compared to average practice technologies. Big gas furnaces for thermal treatment, developed recently by IPSEN industries, Paris, are used in the Department of Germevilliers. Energy recovering burners allow a 40% lower use of gas in cementing, tempering and carbonization.

11.3.1.3 Building Materials Industry

The fuel saving potential in the building materials industry is considerable. We illustrate this claim by reference to recent innovations by PREMACO, at Saint-Mere-Eglise, France, where heat pumps developed by J. Papallardo of Electricité de France, have been installed recently. These pumps almost halve the energy consumption for the industrial drying of gypsum wall boards. Special glasses may be produced by a highly energy saving 'crisver' process, using chemical synthesis.

11.3.1.4 Food Processing Industry

Some final examples present evidence of future energy savings in the food processing industry. For instance, Tekivo Oy, of Helsinki, Finland, has a tunnel oven which is in testing operation. This oven saves 50% of the energy compared to the average practice ovens using light fuel oil.

Wiegand Karlsruhe GmbH of Ettlingen, Germany, produces a tubular evaporation apparatus used in milk production, which decreases energy consumption by more than a quarter.

LURGI GmbH, of Frankfurt am Main, Germany, produces a semi-continuous basic material deodoriser, for use in vegetable oil production, which compared to the previous technology reduces energy consumption by 3 MJ per ton of material. At Hamburg, Germany, Krupp Industrietechnik GmbH, has developed a screw press for sunflower seed, which lowers energy costs by some 50%.

11.3.1.5 Simulation Results

We now turn to scenarios that analyze a turn from conventional average practice technologies to best practice technologies in industry.

We stipulate reductions in fuel requirements for a given level of output by 2% p.a. The 'guesstimate' of '2%' is suggested by the engineering information presented above, and is in line with the assumptions of other studies.

Table 11.4 presents the impact of changes in the matrix of fuel use per unit of output, C, given a 2% p.a. increase in energy efficiency for a sample of industries, and in all sectors together.
Thus, our simulation exercises show that current best technologies can yield lower emissions than average practice technologies now in use.

In most industrial processes, energy use is far above the thermodynamic minimum amount of energy. Faber and Wagenhals [1988] guesstimate that the thermodynamic minimum amount of energy is only some five percent of the actual energy requirements on the average.

Yet, we cannot expect these limits to be reached, even in the very long run. For example, experts of the U.S. Department of Energy, Office of Fossil Energy, expect that in the distant future, fossil energy based

CO_2 Emission Changes (% p.a.)	Germany	UK
Electricity generation	-0.60	-0.72
Iron and steel industry	-0.03	-0.08
Building materials	-0.04	-0.03
Food processing industry	-0.03	-0.03
Chemical products industry	-0.10	-0.08
All sectors	-1.42	-1.38

Table 11.4. CO_2 emission reductions by changing from average to best practice technology: Changes in C.

magnetohydrodynamics and fuel cells will be obtainable with thermodynamic efficiencies up to 60%, a number that is still far from the thermodynamic optimum.

11.3.2 Efficiency of Direct Final Demand Fuel Use

Advanced energy efficient technologies may not only be used in industrial applications. Let us consider three examples:

(1) housing insulation

(2) using more energy-efficient household appliances

(3) increasing the share of district heating.

11.3.2.1 Housing Insulation

Let us assume an improvement of housing insulation from 250 KWh/m² to 150 KWh/m², in an average apartment with an area of 100 m². This yields an energy reduction of 10,000 KWh per apartment. Given our specific emission factor of 0.26 kg CO_2/KWh (assuming oil heating), we would require 2.6 tonnes less of CO_2 per year, per average apartment.

The German Umweltbundesamt [1988] estimates that 95 M tonnes of CO_2 per year (14%) could be saved by switching to state of the art insulation in private buildings. This is a technological upper limit, and our actual estimate is lower. Yet the lesson to be learned is that large amounts of CO_2 emissions can be avoided by turning to improved housing insulation and space heating (especially by using condensing fuel boilers).

11.3.2.2 Household Appliances

As an example, we consider the use of better refrigerators. If an average refrigerator, using 400 kWh per year, is substituted by a state of the art refrigerator using 240 KWh per year, energy savings add up to 160 KWh per year, or CO_2 savings of 42kg of CO_2 per year.

The analysis of a switch to modern washing machines, or other electrical gadgets, yields similar results (see e.g. Khazzoom [1987]). However, an increasing number of appliances and larger gadgets can offset gains made through efficiency improvements (see Lee and Ryu [1991]).

11.3.2.3 District Heating

Increasing the share of district heating also helps to curtail the need for direct heating by fossil fuels. If we assume a substitution of heating oil by district heating, which reduces the electricity demand from fossil-fuel generation by 10%, then CO_2 emissions decline by 10.3 M tonnes in Germany (1.5%) and by 10.1 M tonnes in the UK (1.9%).

However, it is not likely that these savings can be realized in our two countries. In the UK, investment in district heating has been very low, due to cheaper competing energy sources, especially North Sea gas. In Germany, district heating supplies some 8% of all housing units with heat, but further expansion of the district heating system is very limited in scope.

In both countries, combined heat and power schemes for increased district heating are too expensive currently, compared with direct heating, especially using natural gas.

In the future, combined heat and power generation could become more cost-effective, due to increased use of gas in combined cycle turbine plants. However, large investments for distribution networks, and long gestation periods for the development of complete district heating systems, are major obstacles to the further expansion of district heating.

In our survey of potential energy savings originating from improved efficiency of energy use from final demand, we do not present a detailed case study of the impact of increased fuel efficiency of private passenger cars on CO_2 emissions (see Kolb et al. [1989:14-19] for a survey).

11.3.2.4 Simulation Results

This section analyzes the impact of changes in **P**, the matrix of fuel use for direct household final demand, on CO_2 emissions.

Table 11.5 assumes that switches from average to best practice technology results in an annual 2 percent decrease of some or all elements of **P**.

CO_2 Emission Changes (% p.a.)	Germany	UK
Gas only	-0.11	-0.23
Coal only	-0.04	-0.08
Oil only	-0.43	-0.32
Total	-0.58	-0.62

Table 11.5. CO_2 emission reductions by changes from average to best practice technology: Changes in **P** (2% p.a. increase in energy efficiency of final demand).

Both countries show the greatest reduction potential in oil use, and the least reduction possibilities in coal use.

11.3.3 Conclusion

This section showed that many promising advanced energy technologies have reached commercial and pre-commercial stages of development. Their efficiency improvements may help to cut CO_2 emissions. Our results reinforce the outcome of many other assessments (see e.g. most national case studies in the special issue of **Energy Policy**, December 1991). This is that, apart from curtailing energy inputs, the greatest opportunity for abating CO_2 emissions lies in improving energy efficiency.

11.4 Changing the Fuel Mix

The scenarios presented below deal with substitution possibilities between different types of fuel, given the total use of energy in terms of energy content.

We concentrate on natural gas and on non-fossil fuels, as 'clean' substitutes.

11.4.1 Natural Gas

Natural gas supplies about twice as much energy per unit of CO_2 as coal. Therefore, we now look at the upper limits for CO_2 savings from a complete substitution of gas for coal in each sector of the economy.

Broken down by sectors, we get the main percentage CO_2 savings in the electricity sector (total emissions decline by roughly one third). But also the chemical products, construction and some services sectors decrease their emissions significantly by changing their energy supply mix.

Ranking according to CO_2 savings gives the results presented in Table 11.6, which singles out the sectors which are of critical importance in a 'gas for coal' strategy. The industrial sectors which have the highest emission reduction potential include electricity (5), chemical products (11), construction (41), food (38), motor vehicles (23) and mechanical engineering (21).

	Germany		UK	
Rank	Sector	CO_2 Savings (Mt CO_2)	Sector	CO_2 Savings (Mt CO_2)
1	5	26.24	5	21.53
2	11	5.66	42	4.54
3	41	4.18	41	3.89
4	42	4.06	11	2.21
5	23	3.98	38	1.98
6	17	3.96	17	1.80
7	8	3.21	23	1.52
8	38	3.06	21	1.40
9	21	2.80	46	1.23
10	45	2.13	45	1.22

Table 11.6. CO_2 emission reductions by changing the energy mix: Gas for coal in industry (disaggregate results in M tonnes of CO_2).

Total CO_2 savings in industry amount to 11.3% in Germany and 9.8% in the UK. Corresponding CO_2 savings attributable to direct final demand are 0.6% (Germany) and 1.3% (UK).

Next, we look at specific 'gas for other fossil fuels' scenarios, and calculate technical upper limits for CO_2 savings (see Table 11.7).

CO_2 Emission Changes (% p.a.)	Germany	UK
Gas for coal (industry)	-11.33	-9.80
Gas for coal (final demand)	-0.55	-1.27
Gas for oil (industry)	-5.79	-9.57
Gas for oil (final demand)	-5.55	-4.79

Table 11.7. CO_2 emission reductions by changing fuel mix: Substitution of fossil fuels by gas in industry and final demand.

A simulation of a 'gas for coal' strategy in electricity generation shows that CO_2 emissions decline by only 7-8%.

Of course, these upper technological limits cannot be reached without considerable costs. But, independent of cost considerations, the implied decline in CO_2 emissions would not be sufficient to reach the Toronto targets of an overall 20% reduction.

11.4.2 Non-fossil Fuels

Many non-fossil fuels are environmentally desirable as 'clean' sources of energy. To our knowledge, no complete assessment of all economic impacts of non-fossil fuels exists, although many papers have been published dealing with special topics.

For us, comparative studies are most interesting. Schmitt and Junk [1984], as well as Conrad and Henseler-Unger [1986], compare the economic impact of coal-fired versus nuclear power plants in Germany. Hohmeyer [1990] analyzes the costs of wind and photovoltaic versus fossil-fuel and nuclear electricity generation in Germany. He considers private and social electricity costs from a macroeconomic point of view.

Two comprehensive studies by the German Institute of Economic Research and the Fraunhofer Institute for Systems and Innovation Research [DIW/ISI 1984, 1991] assess the potential of renewable energy sources (see also Bölkow et al. [1990]).

Some studies concentrate on solar heat technologies such as Fry [1986], or the papers presented by West and Kreith [1988].

Walton and Hall [1990] describe solar technologies, and identify those that are economic. They compare the private costs of power from solar, wind, nuclear, coal, oil, and gas.

Nitschke [1988] estimates costs of energy from wind converters.

Many studies concentrate on the economic aspects of the nuclear power industry. Chapman [1990] looks at the relationship between the evolution of the nuclear power industry and the costs of nuclear power waste storage. Harding [1990] surveys trends and conclusions in studies of nuclear reactor safety, and their implications for nuclear regulation. Heinze [1989] analyzes the economics of nuclear power plants as they age. Berry and Loudenslager [1987] deal with the impact of nuclear power plant construction on capital costs.

The literature quoted above presents ample results to show that a switch to non-fossil energy systems can make use of many options, including nuclear, solar, hydropower, wind, geothermal, biomass and other methods of energy generation. Before we turn to general simulations assessing the impact of switching the energy mix to non-fossil fuels, let us look at these non-fossil fuels one at a time.

11.4.2.1 Nuclear Energy

The greater exploitation of nuclear energy is an option that can reduce CO_2 emissions. We analyze the contribution that nuclear energy generation currently makes towards the reduction of CO_2 emissions and look at a switch from nuclear energy use to fossil fuel use. Thus, we look at CO_2 emissions which are being avoided by using nuclear energy generation in Germany and the UK.

More specifically, we ask: "What happens if all nuclear fuels are replaced by fossil fuels (i.e gas, coal or oil) in electricity generation?" As we can expect, emissions in both countries increase (see Table 11.8). The lower impact in the UK is due to its lower share of nuclear fuels in electricity generation.

	Change in CO_2 Emissions (% p.a.)	
Percent	Germany	UK
Gas	7.31	1.67
Coal	10.35	2.52
Oil	9.85	2.40

Table 11.8. CO_2 emission changes by substituting fossil fuels for nuclear fuels in energy generation.

In spite of nuclear energy being 'clean', in terms of CO_2 emissions, pollution abatement by using more nuclear energy can cause additional CO_2 pollution, because the construction of nuclear plants requires inputs whose production will generate additional CO_2 emissions. (Cropper [1980] analyzes pollution aspects of nuclear energy use.)

11.4.2.2 Conclusion

Our experiments show that nuclear energy plays an important role in avoiding CO_2 emissions from electricity generation in both countries. However, given the lack of public confidence in nuclear safety, and the current low prices of gas and oil compared to the late 1970s, we do not expect considerable further increases in the nuclear share to satisfy energy demand.

In the next centuries, thermonuclear fusion has the potential to become a major source of 'clean' energy. As we do not expect it to be commercially feasible before the end of the next century, we do not consider it in our simulation exercises.

11.4.2.3 Renewable Energy Sources

We consider energy from renewables to be most environmentally favourable. Unfortunately, there are serious obstacles which impede the widespread use of this energy source now.

Below we present a brief survey of the potential of renewable energy sources for our countries. In a few examples we compare them to an average hard coal power plant with an efficiency of 36% for medium load operation, assuming a specific emission factor of 0.65 kg CO_2 per KWh electricity.

(a) Wind

Wind energy converters require windy and exposed locations. Visual intrusion may be the most important drawback of a large-scale utilization of wind power.

In Germany the share of wind energy in the total electricity production is less than 1%. Efforts are being made to increase this share [Eissenbeiss et al. 1989]. For example, near Niebüll (Northern Germany), in September 1991 Germany's largest 'windpark' was opened. In its final development stage it is planned to supply 27 GWh per year. If the same amount of energy were supplied using hard coal, CO_2 emissions would rise more than 17 Ktonnes. Yet, due to poor wind regimes in most parts of Germany, no substantial rise in wind energy is to be expected in the future.

In the UK, wind power is the most important renewable source of energy [Page et al. 1989]. It contributes 2% of the UK electricity generation. According to Musgrove [1990], wind could satisfy more than 20% of UK energy demand, if off-shore wind converters become cost-effective.

(b) Solar energy

Electricity can be produced with solar technologies, e.g. with solar thermal technology (like concentrating collectors or heliostat technology), which is available in pilot systems, or by photovoltaic systems, where small systems (in the kW range) are already commercially available.

But current techniques are generally more expensive than conventional electricity generation devices, despite advances in photovoltaic energy systems. Also, solar energy conversion capacities in Germany and the UK will not be significant during this century. Therefore we do not expect substantial CO_2 reductions due to the use of solar energy.

(c) Hydropower

The potential of hydropower utilization to produce electricity has been proven on a large scale. A recent unpublished study of the Vereinigung Deutscher Elektrizitätswerke suggests that, by the beginning of the next century, electricity supply from hydropower sources could be increased by 2 billion kWh. But neither in Germany nor in the UK is there a significant potential for long-term increases in hydroelectric generation by building more dams and water reservoirs.

(d) Geothermal energy

Haraden [1989] calculates the CO_2 emission rates for fossil fuels and three forms of geothermal energy. He shows that most geothermal resources produce significantly less CO_2 than fossil fuels for a given level of electricity output.

Although we agree with his conclusion, that geothermal energy may contribute to a reduction of CO_2 emissions, in our two countries the utilization of geothermal deposits (e.g. as hot water power, or high-temperature generation systems) is negligible.

(e) Biomass

Energy production from biomass (e.g. ethanol or methanol from wood/vegetable oils) (see e.g. Palz et al. [1985]) is still in its infancy. Also, biomass systems produce significant quantities of solid waste. Thus we do not expect energy from biomass to be economically feasible during the next two decades.

11.4.3 Conclusion

Summing up, the potential long term contribution of renewables in energy supply is high. At present, their use is mainly restricted to small decentralized units, like small wind systems or single house solar heating. Most renewable sources and technologies are not yet cost-effective.

11.4.4 Simulation Results

We analyze a switch of the energy mix to non-fossil, non-CO_2 emitting fuels, by assuming that gas, coal, and oil energy use (i.e. its energy content in TJ) is cut by 1% across the board. Table 11.9 shows that as a result, in both countries CO_2 emissions decrease. This is by some 0.7% in industry, and by some 0.3% in direct final demand. Almost a third of this decrease is due to the changing input mix in electricity generation.

	Change in CO_2 Emissions (% p.a.)	
Percent	Germany	UK
Industry	-0.71	-0.69
Households	-0.29	-0.31

Table 11.9. CO_2 emission reductions by switching the energy mix to non-fossil fuels (1% decrease in gas, coal, oil energy use).

11.5 Trend Extrapolations

In a set of experiments, we also extrapolated historical experience and used it to project future CO_2 emissions, assuming various changes in final demand, energy efficiency, and/or the energy mix.

Given past values of sectoral final demand, the technology coefficients, and of fuel coefficients, it is easy to perform forecasting exercises based on a linear trend regression. Naturally, given the small support for each regression, these forecasts are in no way reliable. They may be just considered as exercises in assessing the behaviour of the model. We performed scenarios based on the following assumptions:

(a) current trends in components of **y** continued

(b) current trends in components of **C** continued

(c) current trends in components of **P** continued

(d) current trends in components of **y**, **C** and **P** continued.

We also projected CO_2 emissions given that the technical coefficients in the **A** matrices or the Leontief Inverses, and of the **Z** matrices, are changed in line with known past trends.

Extrapolations based on linear trends in **y**, **C** and **P** alone, or in combination, always yielded similar results for both countries. We assumed that the current trend continues, and we set equal to zero any coefficients which are forecast to be less than or equal to zero.

As a result, the current trend of decreasing emissions continues in both countries, until the mid 1990s. But then, if nothing is done, and if there is no technical progress, total emissions begin to rise again in both countries, because substitution possibilities are limited in many industries in the long-term, and the potential for inexpensive energy savings through improved maintenance and control has been largely exhausted.

If in Germany the combined trends in the coefficients of **y**, **C** and **P** continue, total emissions do not increase to the pre-1980 levels. Yet, emissions in the UK surpass their early 1980 level.

Accounting for the trends in the elements of **Z** changed the above results only insignificantly. Trend regressions based on the technology matrices or its Leontief inverses did not produce sensible results.

All trend extrapolation exercises were also performed using a log linear trend approach. The qualitative results did not change and the quantitative results changed only slightly.

Summing up, our forecasting exercises show that the process of CO_2 savings has lost momentum during the last few years. If current trends continue, emissions may rise again in the 1990s. Therefore, there is no reason for a complacency, but, as the following 'plausible' scenarios show, there is also no reason for despair.

11.6 A Sequence of Plausible Scenarios

Finally, we present a sequence of scenarios which we consider as plausible, given our experience with data, past trends and engineering information.

Assumptions on sectoral growth rates are shown in Table 11.10.

Industrial sector	Sector	Growth rate (% p.a.)
Agriculture	1	-1.0
Chemicals	11, 13, 14	2.0
Paper and printing	32-34	2.0
Textiles	36, 37	0.0
Food	38, 39	6.0
Construction	41	2.0
Transport	43	0.0
Services	44-47	4.0

Table 11.10. Sectoral growth rate assumptions for the 'plausible scenarios'.

Growth rates in all sectors not mentioned above are assumed to change proportionally such that a 2.0% p.a. growth rate of total GDP is maintained.

The results presented in Table 11.11 are largely self-explanatory.

1. In experiment 1, we assume a change in the structure of the economy as explained in Table 11.10, combined with a 2% p.a. rate of growth of GDP. This scenario yields increases in emissions of more than 1% p.a. in both countries! (Negative Toronto targets are more than surpassed.)

11.6 A Sequence of Plausible Scenarios

	Scenario	CO_2 Emission Changes (% p.a.)	
		Germany	UK
1	y - Structure and level of final demand.	1.15	1.40
2	C - Energy efficiency improvement: 2% p.a. (in all sectors).	-1.42	-1.38
3	1 and 2 together.	-0.29	-0.01
4	P - Energy efficiency improvement: 2% p.a. (in all sectors).	-0.58	-0.62
5	1 & 2 & 4.	-0.87	-0.63
6	Switch to renewables:		
6a	Industry: 0.5% p.a. decrease in fossil-fuel energy use.	-0.36	-0.35
6b	Final Demand: 0.5% p.a. decrease in fossil-fuel energy use.	-0.14	-0.15
7	1 & 2 & 4 & 6.	-1.36	-1.13
8	Loss of nuclear Assumption: Nuclear energy generation phased out slowly such that emissions increase by 1 M tonnes CO_2 p.a.	0.15	0.19
9	1 & 2 & 4 & 6 & 8.	-1.22	-0.94
10	1 & 2 & 4 & 6 & 8 Change in employment.	2.07	2.48

Table 11.11. A sequence of 'plausible' scenarios for CO_2 emission reductions in Germany and the UK.

2. Next we assume an increase of 2% p.a. in the efficiency of fuel use, in all industry sectors, which yields emissions declining by more than 1.3% p.a. in both countries.

3. Combining the simulations (1) and (2) already gives positive CO_2 savings in both countries. Emission reductions are higher in Germany than in the UK.

4. Now we turn to an improvement of energy efficiency of direct final demand fuel use. If we stipulate a 2% p.a. increase in energy efficiency, emissions decline by more than 0.6% p.a. in both countries.

5. Combining (1), (2) and (4) yields total CO_2 reductions that are not yet large enough to reach the Toronto targets.

6. Thus we add a switch to renewable energy sources. We posit that wind, solar and hydropower utilization increases, so as to decrease fossil fuel generated energy use by 0.5% p.a. Therefore, emissions due to industrial sources decrease by some 0.35 percent per annum and emissions due to direct final demand decline by some 0.15% p.a. in both countries.

7. Combination of experiments (1), (2), (4) and (6) yields annual changes in emissions which surpass the Toronto targets considerably, in both countries.

8. Therefore, we check whether we can afford to avoid nuclear fuels and assume a slow phasing out of nuclear energy generation. We posit that emissions due to increased use of fossil-fuels, which make up for the loss in nuclear fuels, do not surpass 1 M tonnes CO_2 per year. This policy increases emissions by almost 0.2% p.a. in both countries.

9. Summing up experiments (1), (2), (4), (6) and (8) we see that the Toronto targets can still be reached in Germany (old Federal States), and almost attained in the UK.

10. Finally, we present the changes in employment for the combined experiment (1), (2), (4), (6) and (8). In both countries employment increases by more than 2% p.a.

11.7 Conclusions

The simulations presented above rely on assumptions which we feel to be plausible, by and large. They show that the Toronto targets may (almost) be achieved if our economies grow, even if nuclear energy generation in phased out (but very slowly).

The task of reaching the Toronto goals appears to be harder for the UK than for Germany, if we refer (as throughout this study) to the old Federal States of Germany only. However, if we account for the reunification, and allow for the many industrial sites in the former GDR using brown coal or brown coal generated energy, then the picture changes, and we obtain an aggregate decrease in emissions of less than 1% p.a. in experiment (9).

Summing up, reaching the Toronto targets in both countries is (almost) possible, even with a slow phasing out of nuclear fuels. However, it needs considerable efforts, especially to improve energy efficiency and to increase the use of renewable energy sources.

12 A 'Minimum Disruption' Approach to Scenario Analysis

12.1 Introduction

In the previous chapter we presented a number of scenarios of economic restructuring, with the assessment of the corresponding reductions in CO_2 emissions. However, although this approach is well-established, it faces two problems.

One problem with scenario analysis is that the only limit on the number of scenarios that can be considered is the analyst's imagination.

The second is that when one makes certain scenario assumptions concerning structural parameters in the model (i.e. y, C, etc.), one cannot be sure that these adjustments will necessarily lead to the target reduction in CO_2 emissions (i.e. approximately 1% p.a. for the Toronto target).

To try to overcome these problems, we offer a new approach to scenario analysis, using the technique of optimisation. This approach gives an objective measure of what is and is not reasonable as a scenario, and hence gives guidance to, and limits upon, the range of scenarios considered.

The structure of the rest of this chapter is as follows. Section 12.2 outlines a 'minimum disruption' approach to scenario analysis. Section 12.3 applies this approach to alterations to the structure of final demand necessary to achieve a target CO_2 reduction. This includes further constraining the changes to give target GDP growth and/or target employment growth. In Section 12.4 the minimum disruption method is applied to the changes in fuel mix and/or fuel efficiency. The approach is then applied to the structure of inter-industry trading in Section 12.5. Finally, in Section 12.6 some conclusions are drawn.

12.2 The 'Minimum Disruption' Approach

Now from our discussion of the elasticities of CO_2 emissions in Chapter 8, it is clear that we can identify three main areas of changing economic structure that might give rise to reductions in CO_2 emissions. These are:

1. Changes to final demand (y). These are reflected in the elasticities $\varepsilon^C_{y_i}$.

2. Changes to the efficiency of fuel use (**C** and **P**), reflected in the elasticities $\varepsilon^C_{C_{if}}$ and $\varepsilon^C_{P_{if}}$.

3. Changes to the structure of inter-industry trading (**A**), reflected in the elasticities $\varepsilon^C_{a_{.j}}$ and $\varepsilon^C_{a_{i.}}$.

We will therefore consider the necessary changes to these factors to allow the Toronto target to be met. However, as discussed above, the vector and matrices **y**, **P**, **C** and **A** can be varied in an infinite number of ways to achieve any given target alterations in CO_2 emission reduction. Our aim is to select only a limited number of economic restructurings for consideration. The five variables we shall consider for variation are **y**, **P**, **C**, $\{a_{.j}\}$ and $\{a_{i.}\}$. (The final two variables are the column and row sums of the **A** matrix.)

We assess the economic restructuring necessary by using an optimisation technique, based on minimising the 'disruption' to the structure of the economy under consideration, when these variables are varied. Here we shall measure disruption to any particular sector as related to the *proportional change* in that variable. Our overall 'disruption function' is defined to be half of the sum of the squares of the proportional changes of the variable being considered, summed over all sectors. So if we consider some variable, z_i, defined over sectors 1 to n, and if the changes made to achieve a certain target are Δz_i, then the disruption function is defined as:

$$D \equiv \frac{1}{2}\sum_{i=1}^{n}\left(\frac{\Delta z_i}{z_i}\right)^2.$$

We specify this function as we are interested to find 'restructurings' of the economy which generate the least 'economic friction'. We therefore make the assumption that there is an equal social cost associated with a 1% change of a certain variable, irrespective of the sector. For example, a 1% change in the final demand for Motor Vehicles is deemed to be equally disruptive with a 1% change in the final demand for Office Machines.

The minimisation of this disruption function is subject to two types of constraint.

1. The target alteration in CO_2 emissions attributable to altering that variable. (This constraint is most easily expressed in terms of the corresponding elasticity.)

2. Other constraints that seem economically realistic. For example, we may wish to constrain changes in final demand to be consistent with a certain rate of growth of GDP.

As we shall be constraining a quadratic objective function with linear constraints, our problem will be one of quadratic programming [Zoutendjik 1976:Ch. 15].

12.3 Minimising Disruption of Final Demand

We begin by considering the scenarios of changing final demand which achieve a certain target reduction in CO_2 emissions, *ceteris paribus*. We aim to reduce CO_2 emissions with the minimum disruption to the structure of final demand. Recalling that for sector i, final demand is y_i, we define our disruption function in this case as:

$$D \equiv \frac{1}{2} \sum_{i=1}^{n} \left(\frac{\Delta y_i}{y_i} \right)^2. \tag{12.1}$$

This is the square of the proportional change for each sector, summed over all n sectors, times one-half. (The one-half is for later analytical tractability.)

12.3.1 The Constraint on CO_2 Emission Reduction

We now need to introduce the constraint that a certain target reduction of CO_2 emissions is achieved. To express this constraint, we recall that in Section 8.5.1 we defined the elasticity of CO_2 emissions with respect to final demand for output from sector i as:

$$\varepsilon_{y_i}^c \equiv \frac{\partial C/\partial y_i}{C/y_i}. \tag{8.55}$$

Here:

$C \equiv$ total CO_2 emission; $y_i \equiv$ final demand by sector i.

In difference form (8.55) becomes on rearrangement:

$$\left(\frac{\Delta C}{C}\right)_i = \varepsilon_{y_i}^c \frac{\Delta y_i}{y_i}.$$

Summing over all sectors, the total proportional change in CO_2 emissions attributable to changing final demand can be written as:

$$\frac{\Delta C}{C} = \sum_i \varepsilon_{y_i}^c \frac{\Delta y_i}{y_i}. \tag{12.2}$$

We take as a given the required proportional reduction in CO_2 emission attributable to changing the final demand vector **y**. We define this as:

$$R_Y \equiv \frac{\Delta C}{C}. \tag{12.3}$$

Combining (12.2) and (12.3), we seek to satisfy the constraint:

$$\sum_i \varepsilon_{y_i}^c \frac{\Delta y_i}{y_i} = R_Y. \tag{12.4}$$

12.3.2 Minimising y Disruption Subject to CO_2 Emission Target

We seek to minimise the disruption to final demand (12.1), while attaining a certain target proportional reduction in CO_2 emission (12.4). We therefore have the following quadratic programming problem:

$$\min_{\left\{\frac{\Delta y_i}{y_i}\right\}} D = \frac{1}{2} \sum_i \left(\frac{\Delta y_i}{y_i}\right)^2$$

$$\text{subject to:} \quad \sum_i \varepsilon_{y_i}^c \frac{\Delta y_i}{y_i} = R_Y.$$

To simplify the algebra, we define:

12.3.2 Minimising y Disruption Subject to CO_2 Emission Target

$$\hat{y}_i \equiv \frac{\Delta y_i}{y_i}.$$

Here $\hat{y}_i$ is the proportional rate of change of y_i. So our constrained optimisation problem now becomes:

$$\min_{\{\hat{y}_i\}} \quad D = \frac{1}{2}\sum_i \hat{y}_i^2$$

subject to: $\sum_i \varepsilon_{y_i}^C \hat{y}_i = R_Y.$

We solve this constrained optimisation problem with the method of Lagrange [Baumol 1977:Ch. 4]. We first set up the Lagrangian, introducing the Lagrange multiplier λ:

$$L = \frac{1}{2}\sum_{i=1}^n \hat{y}_i^2 + \lambda\left[R_Y - \sum_{i=1}^n \varepsilon_{y_i}^C \hat{y}_i\right]. \tag{12.5}$$

The first order conditions for a minimum are found by differentiating (12.5):

$$\frac{\partial L}{\partial \hat{y}_i} = \hat{y}_i - \lambda\varepsilon_{y_i}^C = 0 \quad i = 1,\ldots,n \tag{12.6}$$

$$\frac{\partial L}{\partial \lambda} = R_Y - \sum_i \varepsilon_{y_i}^C \hat{y}_i = 0. \tag{12.7}$$

There are therefore $(n+1)$ equations in $(n+1)$ unknowns (the $\{\hat{y}_i\}$ and λ). Rearrangement of (12.6) and (12.7) gives:

$$\hat{y}_i = \lambda\varepsilon_{y_i}^C \quad i = 1,\ldots,n \tag{12.8}$$

$$R_Y = \sum_i \varepsilon_{y_i}^C \hat{y}_i. \tag{12.9}$$

Substitution of (12.8) into (12.9) gives:

$$R_Y = \lambda \sum_i \varepsilon_{y_i}^{c2} \qquad i = 1, \ldots, n.$$

Hence we obtain, by further substitution in (12.8):

$$\hat{y}_i = \frac{R_Y \varepsilon_{y_i}^C}{\sum_i \varepsilon_{y_i}^{C2}} \qquad i = 1, \ldots, n.$$

Thus the problem of minimum disruption changes to final demand, to achieve a given proportional change in CO_2 emission, can be solved quite easily. We note that R_Y is negative (a reduction in CO_2 emissions is required), so all of the $\hat{y}_i$ are negative. That is, all of the sectors reduce their outputs. It should also be noted that the reduction in y for sector i is proportional to the elasticity of CO_2 emissions with y for that sector.

12.3.3 Subject to CO_2 Emission Target and GDP Growth Target

It is very likely that, in considering the implications of meeting reductions in emission targets, decision makers will also be concerned with the maintenance of economic growth. Here, by 'economic growth' we shall mean the growth in total GDP (Y). In our input-output formalism, this is simply the sum of the final demands for all sectors, i.e.:

$$Y \equiv \sum_i y_i.$$

So if we wish to maintain a certain rate of economic growth, g, then we require:

$$g = \frac{\Delta Y}{Y} = \frac{\sum_i \Delta y_i}{Y}.$$

For ease of manipulation, we recast this in terms of an elasticity. We define:

$$\varepsilon_{y_i}^Y \equiv \frac{\partial Y / \partial y_i}{Y / y_i}. \qquad (12.10)$$

12.3.3 Subject to CO$_2$ Emission Target and GDP Growth Target

Now:

$$\frac{\partial Y}{\partial y_i} = \frac{\partial \sum_j y_j}{\partial y_i} = 1. \qquad (12.11)$$

Substituting (12.11) in (12.10) gives:

$$\varepsilon^Y_{y_i} = \frac{1}{Y/y_i} = \frac{y_i}{Y}.$$

This elasticity is easily calculated; it is simply the share of sector i in GDP. Now rearrangement of (12.10), and expression in difference form, gives:

$$\left(\frac{\Delta Y}{Y}\right)_i = \varepsilon^Y_{y_i}\left(\frac{\Delta y_i}{y_i}\right) = \varepsilon^Y_{y_i}\hat{y}_i. \qquad (12.12)$$

Summing (12.12) over all sectors, the total proportional growth in GDP is:

$$\frac{\Delta Y}{Y} = \sum_i \varepsilon^Y_{y_i}\hat{y}_i.$$

Now we wish to constrain the proportional rate of growth of GDP to be g. Our constraint is therefore:

$$\sum_i \varepsilon^Y_{y_i}\hat{y}_i = g.$$

So if we wish to attain both a CO$_2$ emission reduction target *and* a growth target, while minimising the disruption to final demand, then we have the quadratic programming problem:

12 A 'Minimum Disruption' Approach to Scenario Analysis

$$\min_{\{\hat{y}_i\}} \frac{1}{2}\sum_i \hat{y}_i^2$$

subject to: $\sum_i \varepsilon^C_{y_i}\hat{y}_i = R_Y$

$$\sum_i \varepsilon^Y_{y_i}\hat{y}_i = g.$$

Introducing the Lagrange multipliers λ and μ, this problem has the Lagrangian:

$$L = \frac{1}{2}\sum_i \hat{y}_i^2 + \lambda\left[R_Y - \sum_i \varepsilon^C_{y_i}\hat{y}_i\right] + \mu\left[g - \sum_i \varepsilon^Y_{y_i}\hat{y}_i\right]. \tag{12.13}$$

Differentiating (12.13) gives the first order conditions for a minimum as:

$$\frac{\partial L}{\partial \hat{y}_i} = \hat{y}_i - \lambda\varepsilon^C_{y_i} - \mu\varepsilon^Y_{y_i} = 0 \quad i = 1,\ldots,n \tag{12.14}$$

$$\frac{\partial L}{\partial \lambda} = R_Y - \sum_i \varepsilon^C_{y_i}\hat{y}_i = 0 \tag{12.15}$$

$$\frac{\partial L}{\partial \mu} = g - \sum_i \varepsilon^Y_{y_i}\hat{y}_i = 0. \tag{12.16}$$

Reorganising (12.14)-(12.16) gives:

$$\hat{y}_i = \lambda\varepsilon^C_{y_i} + \mu\varepsilon^Y_{y_i} \quad i = 1,\ldots,n \tag{12.17}$$

$$R_Y = \sum_i \varepsilon^C_{y_i}\hat{y}_i \tag{12.18}$$

$$g = \sum_i \varepsilon^Y_{y_i}\hat{y}_i. \tag{12.19}$$

Substituting for $\hat{y}_i$ from (12.17) into (12.18) and (12.19) gives:

$$R_Y = \lambda \sum_i \varepsilon_{y_i}^{C^2} + \mu \sum_i \varepsilon_{y_i}^C \varepsilon_{y_i}^Y \qquad (12.20)$$

$$g = \lambda \sum_i \varepsilon_{y_i}^C \varepsilon_{y_i}^Y + \mu \sum_i \varepsilon_{y_i}^{Y^2}. \qquad (12.21)$$

We can solve for λ and μ in (12.20) and (12.21) by introducing the following definitions:

$$\theta_{CC} \equiv \sum_i \varepsilon_{y_i}^{C^2}; \quad \theta_{CY} \equiv \sum_i \varepsilon_{y_i}^C \varepsilon_{y_i}^Y; \quad \theta_{YY} \equiv \sum_i \varepsilon_{y_i}^{Y^2}.$$

These allow us to write (12.20) and (12.21) in matrix form as:

$$\begin{pmatrix} R_Y \\ g \end{pmatrix} = \begin{pmatrix} \theta_{CC} & \theta_{CY} \\ \theta_{CY} & \theta_{YY} \end{pmatrix} \begin{pmatrix} \lambda \\ \mu \end{pmatrix}. \qquad (12.22)$$

We can solve (12.22) by matrix inversion, to give λ and μ as:

$$\begin{pmatrix} \lambda \\ \mu \end{pmatrix} = \begin{pmatrix} \theta_{CC} & \theta_{CY} \\ \theta_{CY} & \theta_{YY} \end{pmatrix}^{-1} \begin{pmatrix} R_Y \\ g \end{pmatrix}.$$

Once λ and μ have been found, it is straightforward to find the $\{\hat{y}_i\}$, by substitution in (12.17). In this case, as we also require aggregate growth of GDP, some of the $\hat{y}_i$ will be negative, and some positive.

12.3.4 Subject to CO_2 Emission Target and Employment Target

An alternative to a concern with the rate of growth of GDP might be a concern with the rate of growth of employment. We define:

$L_i \equiv$ employment in sector i

$L \equiv$ total employment

$l_i \equiv L_i/x_i$ i.e. employment per unit total output.

12 A 'Minimum Disruption' Approach to Scenario Analysis

So we have:

$$L = \sum_i L_i = \sum_i l_i x_i = \mathbf{l'x}. \qquad (12.23)$$

Now we recall our basic input-output equation:

$$\mathbf{x} = (\mathbf{I} - \mathbf{A})^{-1}\mathbf{y}. \qquad (12.24)$$

Substituting (12.24) in (12.23) gives:

$$L = \mathbf{l'}(\mathbf{I} - \mathbf{A})^{-1}\mathbf{y}. \qquad (12.25)$$

We now define:

$$\mathbf{l}^{*\prime} \equiv \mathbf{l'}(\mathbf{I} - \mathbf{A})^{-1}. \qquad (12.26)$$

So substituting (12.26) in (12.25) we have:

$$L = \mathbf{l}^{*\prime}\mathbf{y} = \sum_i l_i^* y_i. \qquad (12.27)$$

In line with the earlier analysis, we define an elasticity:

$$\varepsilon_{y_i}^L \equiv \frac{\partial L/\partial y_i}{L/y_i}. \qquad (12.28)$$

From (12.27) and (12.28) we see that:

$$\varepsilon_{y_i}^L \equiv \frac{l_i^* y_i}{L}.$$

If we wish total employment to grow at proportional rate m, then we need:

$$m = \frac{\Delta L}{L} = \sum_i \frac{\Delta L_i}{L} = \sum_i \varepsilon_{y_i}^L \frac{\Delta y_i}{y_i} = \sum_i \varepsilon_{y_i}^L \hat{y}_i.$$

12.3.4 Subject to CO_2 Emission Target and Employment Target

We can introduce this requirement as a constraint in our quadratic programming problem, as follows:

$$\min_{\{\Delta \hat{y}_i\}} \frac{1}{2} \sum_i \hat{y}_i^2$$

subject to: $\sum_i \varepsilon_{y_i}^C \hat{y}_i = R_Y$

$$\sum_i \varepsilon_{y_i}^L \hat{y}_i = m.$$

This has the Lagrangian:

$$L = \frac{1}{2} \sum_i \hat{y}_i^2 + \lambda \left[R_Y - \sum_i \varepsilon_{y_i}^C \hat{y}_i \right] + \rho \left[m - \sum_i \varepsilon_{y_i}^L \hat{y}_i \right].$$

The first order conditions for a minimum are:

$$\frac{\partial L}{\partial \hat{y}_i} = \hat{y}_i - \lambda \varepsilon_{y_i}^C - \rho \varepsilon_{y_i}^L = 0 \qquad i = 1, \ldots, n \tag{12.29}$$

$$\frac{\partial L}{\partial \lambda} = R_Y - \sum_i \varepsilon_{y_i}^C \hat{y}_i = 0 \tag{12.30}$$

$$\frac{\partial L}{\partial \mu} = m - \sum_i \varepsilon_{y_i}^L \hat{y}_i = 0. \tag{12.31}$$

We can substitute (12.29) in (12.30) and (12.31), to give:

$$R_Y = \lambda \sum_i \varepsilon_{y_i}^{C2} + \rho \sum_i \varepsilon_{y_i}^C \varepsilon_{y_i}^L$$

$$m = \lambda \sum_i \varepsilon_{y_i}^C \varepsilon_{y_i}^L + \rho \sum_i \varepsilon_{y_i}^{L2}.$$

We extend our notation from Section 12.3.3 with:

$$\theta_{CL} \equiv \sum_i \varepsilon^C_{y_i} \varepsilon^L_{y_i}; \quad \theta_{LL} \equiv \sum_i \varepsilon^{L^2}_{y_i}.$$

This allows us to write:

$$\begin{pmatrix} R_Y \\ m \end{pmatrix} = \begin{pmatrix} \theta_{CC} & \theta_{LC} \\ \theta_{LC} & \theta_{LL} \end{pmatrix} \begin{pmatrix} \lambda \\ \mu \end{pmatrix}.$$

When solved, by matrix inversion, this gives λ and μ. These can then be substituted in (12.29), to solve for the $\{\hat{y}_i\}$.

12.3.5 Subject to CO_2 Emission, GDP and Employment Targets

It is also possible to seek the minimum disruption reorganisation of final demand to meet, simultaneously, target reductions in CO_2 emission, GDP growth and employment growth. In this case our problem is:

$$\min_{\{\hat{y}_i\}} \frac{1}{2} \sum_i \hat{y}_i^2$$

subject to: $\sum_i \varepsilon^C_{y_i} \hat{y}_i = R_Y$

$$\sum_i \varepsilon^Y_{y_i} \hat{y}_i = g$$

$$\sum_i \varepsilon^L_{y_i} \hat{y}_i = m.$$

This has the Lagrangian:

$$L = \sum_i \frac{1}{2} \hat{y}_i^2 + \lambda \left[R_Y - \sum_i \varepsilon^C_{y_i} \hat{y}_i \right] + \mu \left[g - \sum_i \varepsilon^Y_{y_i} \hat{y}_i \right] + \rho \left[m - \sum_i \varepsilon^L_{y_i} \hat{y}_i \right].$$

The first order conditions for a minimum are:

12.3.5 Subject to CO_2 Emission, GDP and Employment Targets

$$\frac{\partial L}{\partial \hat{y}_i} = \hat{y}_i - \lambda \varepsilon^C_{y_i} - \mu \varepsilon^Y_{y_i} - \rho \varepsilon^L_{y_i} = 0 \qquad i = 1, \dots, n$$

$$\frac{\partial L}{\partial \lambda} = R_Y - \sum_i \varepsilon^C_{y_i} \hat{y}_i = 0$$

$$\frac{\partial L}{\partial \mu} = g - \sum_i \varepsilon^Y_{y_i} \hat{y}_i = 0$$

$$\frac{\partial L}{\partial \mu} = m - \sum_i \varepsilon^L_{y_i} \hat{y}_i = 0.$$

These reorganise to give:

$$\hat{y}_i = \lambda \varepsilon^C_{y_i} + \mu \varepsilon^Y_{y_i} + \rho \varepsilon^L_{y_i} \qquad i = 1, \dots, n \tag{12.32}$$

$$R_Y = \sum_i \varepsilon^C_{y_i} \hat{y}_i \tag{12.33}$$

$$g = \sum_i \varepsilon^Y_{y_i} \hat{y}_i \tag{12.34}$$

$$m = \sum_i \varepsilon^L_{y_i} \hat{y}_i. \tag{12.35}$$

Substitution of (12.32) in (12.33)-(12.35), and use of the notation defined in Sections 12.3.3 and 12.3.4, gives:

$$\begin{pmatrix} R_Y \\ g \\ m \end{pmatrix} = \begin{pmatrix} \theta_{CC} & \theta_{CY} & \theta_{CL} \\ \theta_{CY} & \theta_{YY} & \theta_{LY} \\ \theta_{CL} & \theta_{LY} & \theta_{LL} \end{pmatrix} \begin{pmatrix} \lambda \\ \mu \\ \rho \end{pmatrix}.$$

$$\tag{12.36}$$

We can solve (12.36) by matrix inversion, and having found λ, μ and ρ, we can substitute for them in (12.32), to solve for the $\{\hat{y}_i\}$.

It is worth noting that, in general, our solution for $\hat{y}_i$ is as a weighted sum of the elasticities found in the constraints, where the weights are the Lagrange multipliers, λ, μ and ρ. The weights are themselves dependent upon the binding values of these constraints (i.e. R_Y, g and m).

12.4 Minimising Disruption of Fuel Use Coefficients

We can now consider how to achieve a target reduction of CO_2 emissions through altering the direct fuel use coefficients embodied in the matrices **P** and **C**, *ceteris paribus*. Now we recall that **P** has only three non-zero elements, so its treatment is relatively trivial. We therefore concentrate our analysis upon the **C** matrix.

We define our disruption function in this case as:

$$\frac{1}{2}\sum_i \sum_f \left(\frac{\Delta c_{if}}{c_{if}}\right)^2 = \sum_i \sum_f \hat{c}_{if}^2.$$

We have again used the 'hat' notation to represent a proportional change.

The constraint on a target reduction in CO_2 emission through altering **C** can be expressed by recalling the definition of the corresponding elasticity:

$$\varepsilon_{c_{if}}^C \equiv \frac{\partial C/\partial c_{if}}{C/c_{if}}.$$

Reorganising, and summing over all sectors (i) and fuels (f) gives:

$$\frac{\Delta C}{C} = \sum_i \sum_f \varepsilon_{c_{if}}^C \frac{\Delta c_{if}}{c_{if}} = \sum_i \sum_f \varepsilon_{c_{if}}^C \hat{c}_{if}.$$

If we take the target proportional reduction in CO_2 emissions attributable to changes in **C** to be R_C, then the constraint we require is:

$$\sum_i \sum_f \varepsilon_{c_{if}}^C \hat{c}_{if} = R_C.$$

The minimisation problem therefore becomes:

$$\min_{\{\hat{c}_{if}\}} \frac{1}{2} \sum_i \sum_f \hat{c}_{if}^2$$

subject to: $\sum_i \sum_f \varepsilon_{c_{if}}^C \hat{c}_{if} = R_C.$

This has Lagrangian:

$$L = \frac{1}{2} \sum_i \sum_f \hat{c}_{if}^2 + \lambda [R_C - \sum_i \sum_f \varepsilon_{c_{if}}^C \hat{c}_{if}].$$

The first-order conditions for a minimum are:

$$\frac{\partial L}{\partial \hat{c}_{if}} = \hat{c}_{if} - \lambda \varepsilon_{c_{if}}^C = 0 \qquad i = 1, \ldots, n \quad f = 1, 2, 3$$

$$\frac{\partial L}{\partial \lambda} = R_C - \sum_i \sum_f \varepsilon_{c_{if}}^C \hat{c}_{if} = 0.$$

These reorganise to give:

$$\hat{c}_{if} = \frac{R_C \varepsilon_{c_{if}}^C}{\sum_i \sum_f \varepsilon_{c_{if}}^{C\,2}} \qquad i = 1, \ldots, n \quad f = 1, 2, 3.$$

A similar calculation can, of course, be made for fuel use by households, by examining the **P** matrix, and its associated elasticities.

12.4.1 Minimising Disruption of C, with Constant Energy Efficiency

An interesting study would be to seek to reduce CO_2 emissions through pure fuel substitution, with no improvement in energy use efficiency. To set up this problem in our minimum disruption framework, we need to introduce further constraints on the total *energy* use by each sector. We can do this by introducing three coefficients relating energy released in combustion to each fuel type. We therefore define:

$\phi_f \equiv$ energy released per unit of fuel f burnt, $f = 1, 2, 3$.

So if we term the energy consumed by sector i as T_i, then we have:

$$T_i = \sum_f \phi_f c_{if} x_i.$$

We again define an elasticity. In this case we use:

$$\varepsilon_{c_{if}}^{T_i} \equiv \frac{\partial T_i / \partial c_{if}}{T_i / c_{if}}. \tag{12.37}$$

From the definition of T_i we see:

$$\varepsilon_{c_{if}}^{T_i} = \frac{\phi_f x_i}{T_i / c_{if}} = \frac{\phi_f x_i c_{if}}{T_i}.$$

Reorganising (12.37), we see that for a given fuel f, and sector i, we can write:

$$\left(\frac{\Delta T_i}{T_i}\right)_f = \varepsilon_{c_{if}}^{T_i} \frac{\Delta c_{if}}{c_{if}} = \varepsilon_{c_{if}}^{T_i} \hat{c}_{if}.$$

Summing over all fuels, this gives:

$$\frac{\Delta T_i}{T_i} = \sum_f \varepsilon_{c_{if}}^{T_i} \hat{c}_{if} \qquad i = 1, \ldots, n.$$

Now we require no change in the energy use by each sector, so our constraints become:

$$\sum_f \varepsilon_{c_{if}}^{T_i} \hat{c}_{if} = 0 \qquad i = 1, \ldots, n.$$

This allows us to write our minimisation problem as:

12.4.1 Minimising Disruption of C, with Constant Energy Efficiency

$$\min_{\{\hat{c}_{if}\}} \frac{1}{2} \sum_i \sum_f \hat{c}_{if}^2$$

subject to: $\sum_i \sum_f \varepsilon_{c_{if}}^C \hat{c}_{if} = R_C$

$$\sum_f \varepsilon_{c_{if}}^{T_i} \hat{c}_{if} = 0 \quad i = 1, \ldots n.$$

Using Lagrange multipliers λ and $\{\mu_i\}$, this has Lagrangian:

$$L = \frac{1}{2} \sum_i \sum_f \hat{c}_{if}^2 + \lambda [R_C - \sum_i \sum_f \varepsilon_{c_{if}}^C \hat{c}_{if}] - \sum_i \mu_i \sum_f \varepsilon_{c_{if}}^{T_i} \hat{c}_{if}.$$

The first-order conditions for a minimum are:

$$\frac{\partial L}{\partial \hat{c}_{if}} = \hat{c}_{if} - \lambda \varepsilon_{c_{if}}^C - \mu_i \varepsilon_{c_{if}}^{T_i} = 0 \quad i = 1, \ldots, n \quad f = 1, 2, 3$$

$$\frac{\partial L}{\partial \lambda} = R_C - \sum_i \sum_f \varepsilon_{c_{if}}^C \hat{c}_{if} = 0$$

$$\frac{\partial L}{\partial \mu_i} = -\sum_f \varepsilon_{c_{if}}^{T_i} \hat{c}_{if} = 0 \quad i = 1, \ldots, n.$$

Substitution and reorganisation gives:

$$\hat{c}_{if} = \lambda \varepsilon_{c_{if}}^C + \mu_i \varepsilon_{c_{if}}^{T_i} = 0 \quad i = 1, \ldots, n \quad f = 1, 2, 3 \qquad (12.38)$$

$$R_C = \lambda \sum_i \sum_f \varepsilon_{c_{if}}^{C\,2} + \sum_i \mu_i \sum_f \varepsilon_{c_{if}}^C \varepsilon_{c_{if}}^{T_i}$$

$$0 = \lambda \sum_f \varepsilon_{c_{if}}^C \varepsilon_{c_{if}}^{T_i} + \mu_i \sum_f \varepsilon_{c_{if}}^{T_i\,2}.$$

Thence we obtain:

$$\lambda = \frac{R_C}{\theta_{CC} - \sum_i \theta_{CT_i}^2/\theta_{T_iT_i}}$$

$$\mu_i = \frac{-\lambda\theta_{CT_i}}{\theta_{T_iT_i}} \qquad i = 1,\ldots,n.$$

Here we have used the following definitions:

$$\theta_{CC} \equiv \sum_i \sum_f \varepsilon_{c_{if}}^{C\,2}; \quad \theta_{CT_i} \equiv \sum_f \varepsilon_{c_{if}}^{C} \varepsilon_{c_{if}}^{T_i}; \quad \theta_{T_iT_i} \equiv \sum_f \varepsilon_{c_{if}}^{T_i\,2}.$$

All of the elements of λ and μ can be calculated, and substitution in (12.38) then gives the $\{\hat{c}_{if}\}$.

12.5 Minimising Disruption of Inter-Industry Trading

We finally turn to the consideration of changing CO_2 emissions because of changing inter-industry trading; i.e. through changes in the **A** matrix. We recall that from Section 8.5.7 we have the definitions:

$$\varepsilon_{a_{.j}}^{C} \equiv \frac{\partial C/\partial a_{.j}}{C/a_{.j}}; \qquad \varepsilon_{a_{i.}}^{C} \equiv \frac{\partial C/\partial a_{i.}}{C/a_{i.}}. \qquad (8.89)$$

The first elasticity corresponds to changing the use of *inputs* into each sector, the second to changing the use of the *outputs* from each sector.

If we wish to constrain the proportional rate of change of CO_2 emissions, attributable to changing inputs to each sector, to R_{Ac} ('c' as the changes are on the columns), then we can require:

$$\sum_j \varepsilon_{a_{.j}}^{C} \frac{\Delta a_{.j}}{a_{.j}} = \sum_j \varepsilon_{a_{.j}}^{C} \hat{a}_{.j} = R_{Ac}.$$

12.5 Minimising Disruption of Inter-Industry Trading

Here we have again used the 'hat' notation to represent proportional changes. We define our disruption function in this case to be:

$$\frac{1}{2}\sum_j \left(\frac{\Delta a_j}{a_j}\right)^2 = \frac{1}{2}\sum_j \hat{a}_j^2.$$

So we can write our problem as:

$$\min_{\{\hat{a}_j\}} \frac{1}{2}\sum_j \hat{a}_j^2$$

subject to: $\sum_j \varepsilon_{a_j}^C \hat{a}_j = R_{Ac}$.

This has Lagrangian:

$$L = \frac{1}{2}\sum_j \hat{a}_j^2 + \lambda\left[R_{Ac} - \sum_j \varepsilon_{a_j}^C \hat{a}_j\right].$$

The first-order conditions for a minimum are:

$$\frac{\partial L}{\partial \hat{a}_j} = \hat{a}_j - \lambda \varepsilon_{a_j}^C = 0 \qquad j = 1,..,n$$

$$\frac{\partial L}{\partial \lambda} = R_{Ac} - \sum_j \varepsilon_{a_j}^C \hat{a}_j = 0.$$

As for the $\{\hat{y}_i\}$ in Section 12.3.2, these can be solved for $\{\hat{a}_j\}$, to give:

$$\hat{a}_j = \frac{R_{Ac}\varepsilon_{a_j}^C}{\sum_i \varepsilon_{a_j}^{C\,2}} \qquad i = 1,...,n.$$

A similar analysis can be undertaken to obtain the corresponding $\{a_{i.}\}$.

12.6 Conclusions

We are able to constrain the range of scenarios to be considered by restricting ourselves to considering, principally, those scenarios which are consistent with the 'minimum disruption' of the economic structure, while achieving a target reduction in CO_2 emissions. We can also impose further constraints, concerning GDP growth, employment growth and energy efficiency. In the next chapter we use the methods so far developed to examine the prospects for reducing CO_2 emissions by Germany and the UK.

13 'Minimum Disruption' Scenario Simulations

13.1 Introduction

In Chapter 12 we established the theory of 'minimum disruption' scenario analysis. In this chapter we apply this theory to the German (1988) and UK (1984) data.

We shall concentrate our analysis on the reductions in CO_2 emissions derived only from productive activity. We do this as there is little of interest to be derived from this approach by examining the three elements that compose the CO_2 emissions directly attributable to final demand, through the direct consumption of fuels. There is, however, considerable interest in using this technique for exploring how the mix of final demand, energy use, and inter-industry trading might need to alter to achieve target CO_2 emission reductions in production, with 'minimum disruption' to the economic structure.

In the analysis that follows, in each case we seek a 1% p.a. reduction in CO_2 emissions. As we explore three cases, if all three types of restructuring were achieved, then the rate of reduction of CO_2 emissions from production would be 3% p.a., which is far more rapid than is required to meet the 'Toronto target' (as we defined it in Section 1.2.3). The results we obtain should, therefore, be regarded purely as indicators of what sectors would need to make various structural changes, to allow the reduction of CO_2 emissions, subject to various constraints, to be achieved with 'minimum disruption'. We compare the minimum disruption scenario rates of change of the various components with the rates of change actually observed for Germany and the UK.

13.2 Data Sources

In Chapter 12 we derived the theory of this minimum disruption approach to scenario analysis, in terms of elasticities. The formulae for most of these elasticities were derived in Section 8.5, and these can be easily calculated from the data described in Chapter 9.

The only further data required is that for the 'output/employment elasticities', discussed in Sections 12.2.4 and 12.2.5. The calculation of these requires the use of employment data, overall and by sector. These data were derived as follows. For Germany we used the *Stastistisches Jahrbuch* [Statistisches Bundesamt 1991]. For the UK the data were taken from the *Annual Abstract of Statistics* [CSO 1987].

13.3 Changes in Final Demand

We first consider the minimum disruption changes necessary to final demand (the y vector), to achieve a 1% p.a. rate of reduction in CO_2 emission by productive activity, *ceteris paribus*. That is, no consideration is given to changing mix or efficiency of use of fuels, nor of changes to the structure of interindustry trading. Here we apply the theory developed in Section 12.3.

We also assess the impacts on employment of these changes to final demand. This is done by taking the changes required in final demand, using the Leontief inverse to calculate the necessary changes in total output, and applying the known labour/output coefficients for each sector.

13.3.1 Changes in Final Demand with no Other Constraints

In Table 13.1 we show the outcome of this minimum disruption calculation, for Germany and the UK, where there are no GDP growth or employment growth constraints (see Sections 12.3.1, 12.3.2). The figures shown are the required proportional rates of change of final demand ($\hat{y}_i$) and employment ($\hat{L}_i$), in the various sectors, and for the aggregate economy.

Looking first at the aggregate figures, we see that a 1% p.a. rate of reduction of emissions requires roughly half that rate of reduction in aggregate GDP, by this calculation. These figures are -0.57% p.a. for Germany and -0.43% p.a. for the UK. These figures are, in themselves, rather encouraging, indicating that, with no constraints on aggregate growth imposed, the costs of CO_2 reduction are less than the 1% p.a. reduction in GDP that might be initially anticipated.

The effects on aggregate employment are also less than a 1% p.a. reduction, being -0.58% p.a. for Germany and -0.38% p.a. for the UK.

We now turn to the rates of change of final demand in each sector, necessary to produce a 1% p.a. reduction in CO_2 emissions. We see that in this unconstrained case, all sectors are required to reduce final demand. For both Germany and the UK, the largest rate of reduction is for Electricity Distribution (Sector 5), at -2.56% p.a. and -2.47% p.a., respectively.

13.3.1 Changes in Final Demand with no Other Constraints

Sector (% change)	Germany $\hat{y}_i$	$\hat{L}_i$	UK $\hat{y}_i$	$\hat{L}_i$
1 Agriculture	-0.17	-0.64	-0.09	-0.33
2 Forestry & fishing	-0.03	-0.32	-0.01	-0.23
3 Electricity; fossil generation	0.00	-1.38	0.00	-1.47
4 Electricity; other generation	0.00	-1.38	0.00	-1.47
5 Electricity; distribution	-2.56	-1.38	-2.47	-1.47
6 Gas	-0.02	-0.41	-0.05	-0.22
7 Water	-0.01	-0.46	-0.03	-0.32
8 Coal extraction, coke ovens, etc	-0.17	-1.00	-0.05	-0.90
9 Extraction of metalliferous ores	-0.01	-0.41	0.00	-0.35
10 Extraction of mineral oil and gas	0.00	-0.38	-0.04	-0.25
11 Chemical products	-1.19	-0.98	-0.64	-0.56
12 Mineral oil processing	-0.11	-0.39	-0.24	-0.63
13 Processing of plastics	-0.11	-0.39	-0.04	-0.30
14 Rubber products	-0.04	-0.33	-0.03	-0.33
15 Stone, clay, cement	-0.11	-0.64	-0.05	-0.54
16 Glass, ceramic goods	-0.10	-0.35	-0.04	-0.28
17 Iron and steel, steel products	-0.60	-0.59	-0.20	-0.32
18 Non-ferrous metals	-0.13	-0.29	-0.09	-0.19
19 Foundries	-0.02	-0.52	-0.01	-0.31
20 Production of steel etc	-0.09	-0.29	-0.13	-0.30
21 Mechanical engineering	-0.51	-0.52	-0.23	-0.30
22 Office machines	-0.07	-0.08	-0.02	-0.06
23 Motor vehicles	-0.77	-0.76	-0.25	-0.32
24 Shipbuilding	-0.02	-0.09	-0.04	-0.08
25 Aerospace equipment	-0.03	-0.04	-0.08	-0.08
26 Electrical engineering	-0.35	-0.41	-0.20	-0.24
27 Instrument engineering	-0.05	-0.17	-0.02	-0.09
28 Engineers' small tools	-0.17	-0.34	-0.02	-0.07
29 Music instruments, toys, etc	-0.03	-0.06	-0.02	-0.24
30 Timber processing	-0.02	-0.37	-0.02	-0.40
31 Wooden furniture	-0.14	-0.25	-0.07	-0.11
32 Pulp, paper, board	-0.14	-0.32	-0.03	-0.27
33 Paper and board products	-0.06	-0.38	-0.03	-0.32
34 Printing and publishing	-0.02	-0.48	-0.04	-0.33
35 Leather, leather goods, footwear	-0.02	-0.05	-0.02	-0.10
36 Textile goods	-0.16	-0.19	-0.08	-0.20
37 Clothes	-0.08	-0.09	-0.03	-0.12
38 Food	-0.92	-0.86	-0.48	-0.44
39 Drink	-0.12	-0.22	-0.14	-0.17
40 Tobacco	-0.02	-0.05	-0.02	-0.05
41 Construction	-0.85	-0.81	-0.67	-0.65
42 Trade wholesale & retail	-0.83	-0.75	-0.84	-0.76
43 Traffic & transport services	-0.63	-0.63	-0.64	-0.56
44 Telecommunications	-0.05	-0.30	-0.05	-0.29
45 Banking, finance, insurance, etc	-0.39	-0.44	-0.22	-0.38
46 Hotels, catering, etc	-0.28	-0.37	-0.21	-0.21
47 Other services	-0.51	-0.55	-0.07	-0.21
Total	-0.57	-0.58	-0.43	-0.38

Table 13.1. Minimum disruption changes to final demand: 1% reduction in CO_2 emission; Germany and the UK.

250 13 'Minimum Disruption' Scenario Simulations

Looking at the rates of change for both Germany and the UK, we see that these are similar in both size and ranking across the sectors. Reference to the theory of this approach, in Section 12.3 shows that these required changes in final demand are, in fact, proportional to the (direct plus indirect) 'CO_2 intensities' for each sector. Thus those sectors with high CO_2 intensities (in production) will be the sectors required to reduce final demand proportionately the most. If the elasticity of demand for all goods were identical, then this effect would be precisely achieved by a steadily increasing 'carbon tax' on all fuels. Even if the demand elasticities varied, this would still be approximately the outcome of an appropriately sized carbon tax.

13.3.2 Demand Changes with a GDP Growth Constraint

We now turn to the situation where we seek a 1% p.a. reduction in CO_2 emission from productive activity, but subject to a 2% p.a. growth in aggregate GDP (see Section 12.3.3). The results of this calculation, for Germany and the UK, are displayed in Table 13.2.

As we are now constraining the CO_2 reductions to be commensurate with economic growth, some sectors will have negative rates of final demand change, and some positive rates. The actual outcome derives from a weighted sum of the (direct plus indirect) CO_2 intensities, and the proportional contribution to GDP of each sector. We note that the effect upon employment, which is not constrained in this case, is that it is required to grow at 1.92% p.a. in Germany, and 2.21% p.a. in the UK. Of course, such growth in employment is unlikely to occur in the long-run, as we might expect labour productivity to continue to increase in both countries.

Looking at the effects on the individual sectors, we see that the required rate of reduction from Electricity Distribution (5) has increased for both countries, with a particularly large rate of reduction (-8.93% p.a.) being required for Germany. Other sectors where the rate of reduction of final demand in Germany exceeds 1% p.a. are: Chemical Products (11) and Iron and Steel (17). For the UK, the sectors with a high rate of required final demand reduction are: Gas (6), Coal Extraction (8), and Mineral Oil Processing (12). On the other hand, there are many sectors where more than 1% p.a. growth rates are required. For Germany these are: Mechanical Engineering (21), Motor Vehicles (23), Electrical Engineering (1.92), Construction (41), Trade Wholesale Retail (42), Banking, etc (45), and Other Services (47). For the UK, such a high growth rate is required for the following sectors: Extraction of Mineral Oil and Gas (10), Mechanical

13.3.2 Demand Changes with a GDP Growth Constraint

	Sector (% change)	Germany $\hat{y}_i$	$\hat{L}_i$	UK $\hat{y}_i$	$\hat{L}_i$
1	Agriculture	-0.10	0.25	0.41	1.48
2	Forestry & fishing	-0.03	1.85	0.04	1.09
3	Electricity; fossil generation	0.00	-2.87	0.00	-1.14
4	Electricity; other generation	0.00	-2.87	0.00	-1.14
5	Electricity; distribution	-8.93	-2.87	-4.31	-1.14
6	Gas	0.25	0.07	-2.67	-0.58
7	Water	-0.02	2.61	0.04	0.98
8	Coal extraction, coke ovens, etc	-0.59	-1.57	-1.04	-0.28
9	Extraction of metalliferous ores	-0.01	0.05	0.00	1.37
10	Extraction of mineral oil and gas	0.00	0.15	1.68	0.47
11	Chemical products	-1.45	-0.53	0.69	1.01
12	Mineral oil processing	0.67	0.88	-3.94	-1.38
13	Processing of plastics	0.17	0.82	0.11	1.45
14	Rubber products	0.04	0.80	0.05	1.36
15	Stone, clay, cement	-0.21	1.94	0.01	3.03
16	Glass, ceramic goods	-0.15	0.69	0.06	1.39
17	Iron and steel, steel products	-1.25	0.10	-0.13	0.99
18	Non-ferrous metals	-0.12	0.49	0.16	0.72
19	Foundries	-0.01	1.22	0.02	1.43
20	Production of steel etc	0.19	0.44	0.35	1.19
21	Mechanical engineering	1.43	1.36	1.10	1.30
22	Office machines	0.37	0.43	0.33	0.50
23	Motor vehicles	1.63	1.61	0.91	1.31
24	Shipbuilding	0.04	0.08	0.19	0.26
25	Aerospace equipment	0.22	0.25	0.73	0.74
26	Electrical engineering	1.93	1.82	1.54	1.55
27	Instrument engineering	0.34	0.93	0.22	0.60
28	Engineers' small tools	0.22	0.67	0.06	0.22
29	Music instruments, toys, etc	0.14	0.25	0.16	1.41
30	Timber processing	0.00	1.07	0.05	2.24
31	Wooden furniture	0.25	0.67	0.27	0.52
32	Pulp, paper, board	-0.28	0.80	0.01	1.33
33	Paper and board products	0.04	0.86	0.10	1.47
34	Printing and publishing	0.05	2.56	0.31	1.84
35	Leather, leather goods, footwear	0.12	0.23	0.17	0.62
36	Textile goods	0.08	0.34	0.43	1.18
37	Clothes	0.36	0.42	0.43	0.89
38	Food	0.15	0.28	1.96	1.94
39	Drink	0.07	0.49	0.36	0.57
40	Tobacco	0.40	0.47	0.14	0.27
41	Construction	2.79	2.76	3.92	3.83
42	Trade wholesale & retail	3.53	2.83	5.62	4.82
43	Traffic & transport services	-0.60	0.31	0.50	1.41
44	Telecommunications	0.79	1.37	0.52	1.88
45	Banking, finance, insurance, etc	4.68	4.17	2.63	2.66
46	Hotels, catering, etc	0.27	0.76	1.86	1.86
47	Other services	5.83	3.78	1.33	1.87
	Total	2.00	1.92	2.00	2.21

Table 13.2. Minimum disruption changes to final demand: 1% reduction in CO_2 emission by Germany and the UK; subject to 2% growth in GDP.

Engineering (21), Electrical Engineering ((26), Food (38), Construction (41), Trade Wholesale & Retail (42), Banking, etc (45), Hotels, Catering, etc. (47), Other Services (47).

Overall, this minimum disruption restructuring of final demand seems not unreasonable. No extremely rapid rates of change of output by any sector are required, and the effect on employment is that of requiring transfer of workers between sectors, rather than large-scale unemployment.

13.3.3 Demand Changes with an Employment Growth Constraint

In Table 13.3 are shown the results from requiring a 1% p.a. reduction in CO_2 emissions, with the requirement that overall employment be unchanged (see Section 12.3.4). This shows relatively little deviation from the results in Table 13.1, further suggesting that alterations to final demand to satisfy target reductions in CO_2 emissions might be achieved with relatively little effect upon employment.

13.3.4 Demand Changes with GDP and Employment Growth Constraints

Finally, Table 13.4 gives the results for a 1% p.a. reduction in CO_2 emissions, with 2% p.a. growth in GDP *and* employment (see Section 12.3.5). This has been assumed so that the growing productivity of labour can be taken into account, without needing the labour coefficients to be altered. So here we implicitly assume that the 2% p.a. rate of growth of GDP can be accommodated by a constant work force. We see that Table 13.4 closely resembles Table 13.2, suggesting that with not unreasonable structural changes in final demand, considerable reductions in CO_2 emissions are feasible, while *still* maintaining reasonable rates of GDP growth, and without adding to unemployment.

13.3.4 Demand Changes with GDP and Employment Growth Constraints 253

Sector (% change)	Germany $\hat{y}_i$	$\hat{L}_i$	UK $\hat{y}_i$	$\hat{L}_i$
1 Agriculture	-0.08	-0.33	-0.04	-0.14
2 Forestry & fishing	-0.02	0.11	0.00	-0.10
3 Electricity; fossil generation	0.00	-1.71	0.00	-1.15
4 Electricity; other generation	0.00	-1.71	0.00	-1.15
5 Electricity; distribution	-4.02	-1.71	-2.09	-1.15
6 Gas	-0.02	-0.34	-1.26	-0.80
7 Water	-0.01	0.15	-0.02	-0.17
8 Coal extraction, coke ovens, etc	-0.26	-1.13	-0.44	-0.73
9 Extraction of metalliferous ores	-0.02	-0.31	0.00	-0.19
10 Extraction of mineral oil and gas	0.00	-0.30	-0.02	-0.48
11 Chemical products	-1.45	-1.03	-0.48	-0.39
12 Mineral oil processing	-0.15	-0.21	-1.93	-1.21
13 Processing of plastics	-0.06	-0.10	-0.03	-0.13
14 Rubber products	-0.02	-0.05	-0.02	-0.15
15 Stone, clay, cement	-0.13	-0.02	-0.04	-0.24
16 Glass, ceramic goods	-0.11	-0.10	-0.03	-0.13
17 Iron and steel, steel products	-0.82	-0.42	-0.16	-0.19
18 Non-ferrous metals	-0.15	-0.11	-0.07	-0.11
19 Foundries	0.00	-0.04	-0.01	-0.16
20 Production of steel etc	-0.01	-0.11	-0.07	-0.16
21 Mechanical engineering	0.11	0.07	-0.08	-0.15
22 Office machines	0.01	0.01	-0.01	-0.03
23 Motor vehicles	-0.11	-0.11	-0.14	-0.16
24 Shipbuilding	0.00	-0.05	-0.02	-0.08
25 Aerospace equipment	0.01	0.00	-0.02	-0.02
26 Electrical engineering	0.19	0.12	-0.06	-0.09
27 Instrument engineering	0.06	0.06	0.00	-0.03
28 Engineers' small tools	-0.07	-0.09	-0.01	-0.05
29 Music instruments, toys, etc	0.02	0.02	-0.01	-0.05
30 Timber processing	-0.02	0.00	-0.01	-0.15
31 Wooden furniture	0.02	0.02	-0.04	-0.05
32 Pulp, paper, board	-0.19	-0.10	-0.03	-0.10
33 Paper and board products	-0.07	-0.12	-0.02	-0.13
34 Printing and publishing	-0.01	0.09	-0.01	-0.10
35 Leather, leather goods, footwear	0.02	0.02	0.00	-0.01
36 Textile goods	-0.10	-0.06	-0.04	-0.06
37 Clothes	0.09	0.09	0.01	-0.01
38 Food	-0.53	-0.47	-0.25	-0.20
39 Drink	-0.10	-0.06	-0.09	-0.08
40 Tobacco	-0.01	0.00	-0.01	-0.02
41 Construction	0.05	0.05	-0.30	-0.28
42 Trade wholesale & retail	0.46	0.27	-0.14	-0.15
43 Traffic & transport services	-0.62	-0.40	-0.46	-0.34
44 Telecommunications	0.11	0.05	-0.01	-0.06
45 Banking, finance, insurance, etc	0.55	0.44	-0.03	-0.11
46 Hotels, catering, etc	0.03	0.01	0.21	0.21
47 Other services	0.40	0.17	0.57	0.33
Total	-0.03	0.00	-0.20	0.00

Table 13.3. Minimum disruption changes to final demand: 1% reduction in CO_2 emission by Germany and the UK; subject to no change in employment.

13 'Minimum Disruption' Scenario Simulations

Sector (% change)	Germany $\hat{y}_i$	Germany $\hat{L}_i$	UK $\hat{y}_i$	UK $\hat{L}_i$
1 Agriculture	0.02	0.44	0.41	1.40
2 Forestry & fishing	-0.02	1.76	0.04	1.06
3 Electricity; fossil generation	0.00	-2.98	0.00	-2.96
4 Electricity; other generation	0.00	-2.98	0.00	-2.96
5 Electricity; distribution	-9.09	-2.98	-7.85	-2.96
6 Gas	0.15	0.02	0.42	1.15
7 Water	-0.02	2.47	0.01	0.85
8 Coal extraction, coke ovens, etc	-0.59	-1.60	-0.12	-1.11
9 Extraction of metalliferous ores	-0.01	0.06	0.00	1.31
10 Extraction of mineral oil and gas	0.00	0.09	2.14	1.34
11 Chemical products	-1.79	-0.77	0.24	0.69
12 Mineral oil processing	0.33	0.72	0.19	-0.03
13 Processing of plastics	0.15	0.86	0.09	1.42
14 Rubber products	0.05	0.86	0.03	1.28
15 Stone, clay, cement	-0.22	2.01	-0.04	3.12
16 Glass, ceramic goods	-0.15	0.73	0.02	1.37
17 Iron and steel, steel products	-1.38	0.12	-0.37	0.88
18 Non-ferrous metals	-0.17	0.50	0.11	0.68
19 Foundries	0.02	1.37	0.01	1.42
20 Production of steel etc	0.21	0.48	0.28	1.15
21 Mechanical engineering	1.77	1.64	1.03	1.27
22 Office machines	0.34	0.40	0.40	0.56
23 Motor vehicles	1.86	1.80	0.85	1.24
24 Shipbuilding	0.05	0.09	0.19	0.36
25 Aerospace equipment	0.19	0.23	0.81	0.81
26 Electrical engineering	2.01	1.87	1.66	1.64
27 Instrument engineering	0.38	0.92	0.24	0.61
28 Engineers' small tools	0.24	0.72	0.05	0.21
29 Music instruments, toys, etc	0.15	0.27	0.16	1.38
30 Timber processing	-0.01	1.16	0.05	2.26
31 Wooden furniture	0.38	0.79	0.26	0.51
32 Pulp. paper, board	-0.32	0.76	-0.02	1.29
33 Paper and board products	-0.01	0.85	0.09	1.42
34 Printing and publishing	0.05	2.38	0.34	1.81
35 Leather, leather goods, footwear	0.14	0.24	0.18	0.60
36 Textile goods	0.09	0.37	0.44	1.17
37 Clothes	0.48	0.53	0.47	0.90
38 Food	0.41	0.52	1.87	1.84
39 Drink	0.05	0.50	0.29	0.48
40 Tobacco	0.27	0.36	0.15	0.27
41 Construction	3.00	2.84	4.06	3.95
42 Trade wholesale & retail	4.11	3.16	5.59	4.78
43 Traffic & transport services	-0.59	0.35	-0.05	1.15
44 Telecommunications	0.74	1.35	0.58	1.87
45 Banking, finance, insurance, etc	4.43	3.84	2.96	2.78
46 Hotels, catering, etc	0.59	0.98	1.56	1.56
47 Other services	5.09	3.33	0.54	1.37
Total	2.00	2.00	2.00	2.00

Table 13.4. Minimum disruption changes to final demand: 1% reduction in CO_2 emission by Germany and the UK; subject to 2% growth in GDP and employment.

In Table 13.5 we show the annualised rates of change of final demand for each sector, for Germany and the UK. Comparing Tables 13.5 with Table 13.2, the outcome for 2% growth p.a. (no employment constraint), we see that the minimum disruption rates of change of final demand to achieve 1% p.a. reduction in CO_2 emissions are rather different from those actually achieved in the relatively recent past. However, there are some hopeful signs. The rate of change of final demand for Coal Extraction (8) has been markedly more negative than that required by the minimum disruption analysis. Also, the actual relatively high rates of growth in the services sectors (44-47) are also found in the minimum disruption case. The clearest divergence is in Electricity Distribution (5). The high rates of reduction required in the minimum disruption case are not seen in reality, with observed growth rates of 1.81% p.a. for Germany and 1.16% p.a. for the UK.

The minimum disruption analysis has indicated the structural change to final demand that would be necessary to achieve a target reduction in CO_2 emissions. How such structural change could be attained is, of course, a separate issue. We noted above that, with no constraints on growth and employment, something akin to such restructuring could result from a carbon tax. To also achieve economic growth and steady employment would require the use of further economic policy instruments, and it is not our aim in this study to deal with the precise nature of such instruments, nor how they should be invoked. However, such instruments could include employment taxes and subsidies, as are already employed as part of industrial and regional policies, throughout the EC. Further adjustment could also be achieved by use of 'more-than-proportional' carbon taxes, so that the bulk of the impact would fall on commodities which are most CO_2 intensive, such as fossil fuel generated electricity.

13.4 Changes in Fuel Mix and Fuel Efficiency

We now turn to the minimum disruption changes necessary to the fuel mix, and the efficiency of energy use, to reduce CO_2 emissions by 1% p.a., *ceteris paribus* (i.e. we consider changes to the C matrix). Here we apply the theory developed in Section 12.4.

Sector (% change)	Germany 78-88	UK 68-84
1 Agriculture	4.48	-2.36
2 Forestry & fishing	4.05	-1.85
3 Electricity; fossil generation	0.00	0.00
4 Electricity; other generation	0.00	0.00
5 Electricity; distribution	1.81	1.16
6 Gas	4.81	5.11
7 Water	0.18	-3.67
8 Coal extraction, coke ovens, etc	-13.33	-12.18
9 Extraction of metalliferous ores	8.89	-13.47
10 Extraction of mineral oil and gas	-0.43	60.71
11 Chemical products	4.05	3.76
12 Mineral oil processing	-1.73	7.36
13 Processing of plastics	7.22	10.46
14 Rubber products	1.90	-0.03
15 Stone, clay, cement	0.92	1.49
16 Glass, ceramic goods	2.49	-0.38
17 Iron and steel, steel products	2.15	-0.10
18 Non-ferrous metals	2.58	3.97
19 Foundries	4.46	12.27
20 Production of steel etc	0.13	-2.81
21 Mechanical engineering	1.28	-1.05
22 Office machines	10.96	10.70
23 Motor vehicles	2.92	-0.44
24 Shipbuilding	-4.11	-1.66
25 Aerospace equipment	6.76	4.00
26 Electrical engineering	4.14	2.48
27 Instrument engineering	2.50	0.23
28 Engineers' small tools	2.77	9.62
29 Music instruments, toys, etc	-1.32	-1.48
30 Timber processing	6.47	-2.30
31 Wooden furniture	-1.17	2.37
32 Pulp, paper, board	8.45	1.67
33 Paper and board products	4.74	2.15
34 Printing and publishing	4.66	-0.73
35 Leather, leather goods, footwear	-0.70	-1.67
36 Textile goods	0.59	-1.66
37 Clothes	-1.55	-0.32
38 Food	1.46	0.72
39 Drink	0.50	3.27
40 Tobacco	-0.15	-1.11
41 Construction	0.55	-0.75
42 Trade wholesale & retail	2.11	2.58
43 Traffic & transport services	2.32	-1.15
44 Telecommunications	5.87	1.45
45 Banking, finance, insurance, etc	3.19	2.69
46 Hotels, catering, etc	2.26	1.48
47 Other services	5.17	0.05
Total	2.46	2.07

Table 13.5. Actual annualised rates of change of final demand: Germany (1978-88) and UK (1968-84).

13.4.1 Change in Fuel Mix with no Constraints

We begin by allowing all coefficients of the C matrix to vary, without constraint (see Section 12.4). The outcome of this calculation is shown in Table 13.6 for Germany, and Table 13.7 for the UK. We see that all coefficients show negative rates of change, with the largest rates of reduction being for the Coal coefficients, and the smallest for the Gas coefficients. This reflects the higher CO_2 intensity of coal as compared with oil and gas. Much the biggest calculated rate of change, for both Germany and the UK, is in the use of coal by Electricity, Fossil Generation (3). The required rates of reduction are 3.23% p.a. for Germany and 1.76% p.a. for the UK. Both these are rather large, but as the discussion in Section 11.3.1.1 indicates, not beyond the bounds of possibility. Further, recall that these improvements in fuel efficiency would, in themselves, be sufficient to meet the 'Toronto Target' for the production section of these economies. Clearly, there is ample scope for improvement in fuel efficiency, without making outrageous demands upon any particular sector.

13.4.2 Change in Fuel Mix with Constant Energy Efficiency Constraint

As well as improvements in overall fuel efficiency, we might seek reductions of CO_2 emissions by pure fuel substitution, with the *energy* efficiency of each sector unchanged (see Section 12.4.1). The results of this calculation are shown in Tables 13.8 (Germany) and 13.9 (UK).

The results here are striking. The minimum disruption analysis suggests massive and rapid shifts from the use of coal and oil, towards the use of gas, by all sectors except Electricity; Fossil Generation (3). In this sector there is also substitution away from coal towards *oil*. Fortunately, such a shift is already underway, particularly in the UK. This becomes particularly apparent when we look at Tables 13.10 and 13.11. Here are shown the actual proportional rates of change of the elements of the C matrix for Germany (1978-88) and the UK (1968-84). For the UK, these correspond reasonably well with the minimum disruption results, except that the observed rates of change are larger than those needed for only a 1% p.a. rate of reduction in CO_2 emissions, *ceteris paribus*. In the UK the shift to gas has already begun. In Germany, however, which does not have such large indigenous gas resources, the shift has been much slower. It is also worth noting that, in aggregate, the UK has also achieved an overall improvement in energy efficiency (-2.50% p.a.), while for Germany, energy efficiency in production has been worsening (0.85% p.a.).

13 'Minimum Disruption' Scenario Simulations

Germany Sector (% change)	Gas	Coal	Oil
1 Agriculture	-0.00	-0.00	-0.26
2 Forestry & fishing	-0.01	-0.00	-0.04
3 Electricity; fossil generation	-0.36	-3.23	-0.22
4 Electricity; other generation	0.00	0.00	0.00
5 Electricity; distribution	0.00	0.00	0.00
6 Gas	0.00	0.00	-0.00
7 Water	0.00	0.00	-0.00
8 Coal extraction, coke ovens, etc	-0.01	-1.30	-0.00
9 Extraction of metalliferous ores	-0.01	-0.00	-0.00
10 Extraction of mineral oil and gas	-0.00	0.00	-0.00
11 Chemical products	-0.32	-0.11	-0.27
12 Mineral oil processing	-0.02	0.00	-0.08
13 Processing of plastics	-0.01	-0.00	-0.01
14 Rubber products	-0.01	-0.00	-0.01
15 Stone, clay, cement	-0.06	-0.09	-0.07
16 Glass, ceramic goods	-0.06	-0.00	-0.03
17 Iron and steel, steel products	-0.15	-0.06	-0.07
18 Non-ferrous metals	-0.03	-0.00	-0.01
19 Foundries	-0.02	-0.00	-0.01
20 Production of steel etc	-0.00	-0.00	-0.01
21 Mechanical engineering	-0.02	-0.00	-0.06
22 Office machines	-0.00	0.00	-0.00
23 Motor vehicles	-0.04	-0.00	-0.03
24 Shipbuilding	-0.00	0.00	-0.00
25 Aerospace equipment	-0.00	-0.00	-0.00
26 Electrical engineering	-0.01	-0.00	-0.04
27 Instrument engineering	-0.00	0.00	-0.01
28 Engineers' small tools	-0.01	-0.00	-0.02
29 Music instruments, toys, etc	-0.00	0.00	-0.00
30 Timber processing	-0.00	-0.00	-0.02
31 Wooden furniture	-0.00	-0.00	-0.02
32 Pulp, paper, board	-0.04	-0.02	-0.04
33 Paper and board products	-0.01	0.00	-0.01
34 Printing and publishing	-0.01	0.00	-0.01
35 Leather, leather goods, footwear	-0.00	-0.00	-0.00
36 Textile goods	-0.03	-0.00	-0.02
37 Clothes	-0.00	-0.00	-0.01
38 Food	-0.08	-0.01	-0.12
39 Drink	-0.02	-0.00	-0.03
40 Tobacco	-0.00	-0.00	-0.00
41 Construction	-0.00	-0.00	-0.16
42 Trade wholesale & retail	-0.04	-0.00	-0.27
43 Traffic & transport services	-0.00	0.00	-0.67
44 Telecommunications	-0.00	-0.00	-0.02
45 Banking, finance, insurance, etc	-0.01	0.00	-0.03
46 Hotels, catering, etc	-0.02	-0.00	-0.04
47 Other services	-0.03	-0.00	-0.23
Total	-0.19	-2.50	-0.28

Table 13.6. Minimum disruption changes to fuel use (i.e. C): 1% reduction in CO_2 emission by Germany.

13.4.2 Change in Fuel Mix with Constant Energy Efficiency Constraint

Sector (% change)	UK Gas	Coal	Oil
1 Agriculture	-0.00	-0.00	-0.04
2 Forestry & fishing	0.00	0.00	-0.00
3 Electricity; fossil generation	-0.01	-1.76	-0.96
4 Electricity; other generation	0.00	0.00	0.00
5 Electricity; distribution	0.00	0.00	0.00
6 Gas	-0.87	-0.00	-0.01
7 Water	-0.00	0.00	-0.01
8 Coal extraction, coke ovens, etc	-0.00	-0.35	-0.00
9 Extraction of metalliferous ores	-0.00	0.00	-0.01
10 Extraction of mineral oil and gas	0.00	0.00	0.00
11 Chemical products	-0.20	-0.02	-0.08
12 Mineral oil processing	-0.00	-0.00	-1.48
13 Processing of plastics	-0.01	-0.00	-0.03
14 Rubber products	-0.01	-0.00	-0.02
15 Stone, clay, cement	-0.02	-0.07	-0.02
16 Glass, ceramic goods	-0.02	-0.00	-0.01
17 Iron and steel, steel products	-0.05	-0.21	-0.04
18 Non-ferrous metals	-0.01	-0.02	-0.01
19 Foundries	-0.01	-0.00	-0.01
20 Production of steel etc	-0.01	-0.00	-0.01
21 Mechanical engineering	-0.01	-0.00	-0.01
22 Office machines	-0.00	0.00	-0.00
23 Motor vehicles	-0.02	-0.01	-0.01
24 Shipbuilding	-0.00	0.00	-0.00
25 Aerospace equipment	-0.00	-0.00	-0.00
26 Electrical engineering	-0.01	-0.00	-0.01
27 Instrument engineering	-0.00	0.00	-0.00
28 Engineers' small tools	-0.00	0.00	-0.00
29 Music instruments, toys, etc	-0.00	0.00	-0.00
30 Timber processing	-0.00	-0.00	-0.01
31 Wooden furniture	-0.00	-0.00	-0.01
32 Pulp, paper, board	-0.02	-0.02	-0.01
33 Paper and board products	-0.01	-0.00	-0.01
34 Printing and publishing	-0.00	0.00	-0.01
35 Leather, leather goods, footwear	-0.00	-0.00	-0.00
36 Textile goods	-0.01	-0.01	-0.01
37 Clothes	-0.00	-0.00	-0.00
38 Food	-0.04	-0.01	-0.05
39 Drink	-0.01	-0.00	-0.02
40 Tobacco	0.00	0.00	0.00
41 Construction	-0.00	0.00	-0.04
42 Trade wholesale & retail	-0.05	0.00	-0.02
43 Traffic & transport services	0.00	0.00	-0.60
44 Telecommunications	-0.00	0.00	-0.02
45 Banking, finance, insurance, etc	-0.03	0.00	-0.00
46 Hotels, catering, etc	-0.01	-0.00	-0.00
47 Other services	-0.01	0.00	-0.00
Total	-0.10	-1.44	-0.72

Table 13.7. Minimum disruption changes to fuel use (i.e. **C**): 1% reduction in CO_2 emission by the UK.

	Germany		
Sector (% change)	Gas	Coal	Oil
1 Agriculture	0.32	-0.03	-0.00
2 Forestry & fishing	0.64	-0.01	-0.12
3 Electricity; fossil generation	48.03	-7.75	3.14
4 Electricity; other generation	0.00	0.00	0.00
5 Electricity; distribution	0.00	0.00	0.00
6 Gas	0.00	0.00	0.00
7 Water	0.00	0.00	0.00
8 Coal extraction, coke ovens, etc	0.96	-0.01	0.05
9 Extraction of metalliferous ores	0.03	-0.08	-0.16
10 Extraction of mineral oil and gas	0.02	0.00	-0.03
11 Chemical products	11.78	-8.10	-15.74
12 Mineral oil processing	1.56	0.00	-0.40
13 Processing of plastics	0.44	-0.01	-0.56
14 Rubber products	0.19	-0.17	-0.38
15 Stone, clay, cement	4.75	-3.15	-1.42
16 Glass, ceramic goods	1.02	-0.04	-2.41
17 Iron and steel, steel products	2.91	-4.63	-4.92
18 Non-ferrous metals	0.21	-0.40	-0.66
19 Foundries	0.28	-0.04	-0.67
20 Production of steel etc	0.33	-0.01	-0.15
21 Mechanical engineering	1.90	-0.05	-0.97
22 Office machines	0.09	0.00	-0.15
23 Motor vehicles	1.00	-0.32	-1.77
24 Shipbuilding	0.07	0.00	-0.06
25 Aerospace equipment	0.04	-0.02	-0.10
26 Electrical engineering	1.07	-0.03	-0.34
27 Instrument engineering	0.16	0.00	-0.08
28 Engineers' small tools	0.92	-0.01	-0.73
29 Music instruments, toys, etc	0.06	0.00	-0.03
30 Timber processing	0.23	-0.05	-0.03
31 Wooden furniture	0.09	-0.01	-0.00
32 Pulp, paper, board	1.84	-1.23	-1.83
33 Paper and board products	0.32	0.00	-0.41
34 Printing and publishing	0.27	0.00	-0.29
35 Leather, leather goods, footwear	0.09	-0.00	-0.03
36 Textile goods	0.92	-0.28	-1.44
37 Clothes	0.09	-0.00	-0.01
38 Food	5.02	-0.44	-4.60
39 Drink	1.19	-0.05	-1.14
40 Tobacco	0.06	-0.01	-0.04
41 Construction	0.19	-0.03	-0.00
42 Trade wholesale & retail	4.60	-0.02	-0.97
43 Traffic & transport services	0.16	0.00	-0.00
44 Telecommunications	0.19	-0.00	-0.02
45 Banking, finance, insurance, etc	0.74	0.00	-0.26
46 Hotels, catering, etc	1.59	-0.02	-0.91
47 Other services	3.77	-0.01	-0.77
Total	15.93	-5.48	-1.87

Table 13.8. Minimum disruption changes to fuel use (i.e. C): 1% reduction in CO_2 emission by Germany; subject to no change in overall energy efficiency by each sector.

13.4.2 Change in Fuel Mix with Constant Energy Efficiency Constraint

Sector (% change)	UK Gas	Coal	Oil
1 Agriculture	0.66	-0.09	-0.02
2 Forestry & fishing	0.03	0.00	-0.00
3 Electricity; fossil generation	10.62	-30.76	53.63
4 Electricity; other generation	0.00	0.00	0.00
5 Electricity; distribution	0.00	0.00	0.00
6 Gas	0.95	-0.40	-3.15
7 Water	0.23	-0.00	-0.01
8 Coal extraction, coke ovens, etc	0.46	-0.01	0.07
9 Extraction of metalliferous ores	0.41	0.00	-0.05
10 Extraction of mineral oil and gas	0.00	0.00	0.00
11 Chemical products	10.68	-8.42	-33.49
12 Mineral oil processing	0.10	-0.01	-0.00
13 Processing of plastics	4.63	-0.46	-2.25
14 Rubber products	3.25	-0.70	-1.97
15 Stone, clay, cement	13.60	-7.27	-0.96
16 Glass, ceramic goods	0.65	-0.71	-2.74
17 Iron and steel, steel products	34.80	-13.89	0.15
18 Non-ferrous metals	4.56	-4.60	-1.04
19 Foundries	0.91	-1.20	-2.34
20 Production of steel etc	1.73	-0.31	-2.89
21 Mechanical engineering	3.19	-0.48	-4.15
22 Office machines	0.13	0.00	-0.17
23 Motor vehicles	2.80	-2.93	-4.08
24 Shipbuilding	0.55	-0.00	-0.31
25 Aerospace equipment	0.78	-0.63	-1.19
26 Electrical engineering	1.60	-0.29	-3.49
27 Instrument engineering	0.12	0.00	-0.31
28 Engineers' small tools	0.09	0.00	-0.22
29 Music instruments, toys, etc	0.98	-0.01	-0.52
30 Timber processing	0.37	-0.01	-0.02
31 Wooden furniture	1.00	-0.03	-0.16
32 Pulp, paper, board	4.35	-5.45	-2.35
33 Paper and board products	2.14	-0.10	-1.43
34 Printing and publishing	1.42	-0.01	-1.32
35 Leather, leather goods, footwear	0.24	-0.14	-0.33
36 Textile goods	2.94	-1.98	-2.46
37 Clothes	0.43	-0.04	-0.57
38 Food	10.95	-3.15	-11.83
39 Drink	4.16	-0.91	-3.34
40 Tobacco	0.00	0.00	0.00
41 Construction	0.90	0.00	-0.047
42 Trade wholesale & retail	2.57	0.00	-8.31
43 Traffic & transport services	0.00	0.00	0.00
44 Telecommunications	0.25	0.00	-0.73
45 Banking, finance, insurance, etc	0.05	0.00	-1.06
46 Hotels, catering, etc	0.03	-0.49	-0.28
47 Other services	0.10	0.00	-0.59
Total	8.65	-26.33	19.91

Table 13.9. Minimum disruption changes to fuel use (i.e. C): 1% reduction in CO_2 emission by the UK; subject to no change in overall energy efficiency by each sector.

13 'Minimum Disruption' Scenario Simulations

Sector (% change)	Germany Gas	Coal	Oil	Energy
1 Agriculture	3.90	-0.87	-0.98	-0.05
2 Forestry & fishing	5.95	-2.57	-4.76	-2.83
3 Electricity; fossil generation	-5.28	2.13	-6.14	-1.47
4 Electricity; other generation	0.00	0.00	0.00	0.00
5 Electricity; distribution	0.00	0.00	0.00	0.00
6 Gas	0.00	0.00	-16.91	-14.93
7 Water	0.00	0.00	-0.59	-0.35
8 Coal extraction, coke ovens, etc	3.10	-0.27	-11.03	-2.56
9 Extraction of metalliferous ores	-8.32	0.00	-13.89	-3.77
10 Extraction of mineral oil and gas	1.77	0.00	9.95	3.06
11 Chemical products	0.43	3.56	-3.13	1.62
12 Mineral oil processing	2.95	0.00	-8.94	-8.86
13 Processing of plastics	3.68	-12.42	-10.84	-1.04
14 Rubber products	0.87	0.53	-12.00	-3.56
15 Stone, clay, cement	-1.84	11.61	-10.99	-4.74
16 Glass, ceramic goods	-2.15	-1.18	-7.46	-2.68
17 Iron and steel, steel products	-2.76	45.37	-8.44	-3.65
18 Non-ferrous metals	2.15	-7.85	-11.98	-3.52
19 Foundries	-1.54	-6.58	-6.28	-3.31
20 Production of steel etc	-0.19	-5.44	-3.17	-2.86
21 Mechanical engineering	1.45	-6.85	-4.85	-2.79
22 Office machines	-8.09	0.00	-14.09	-2.32
23 Motor vehicles	-1.27	-7.25	-10.32	-3.16
24 Shipbuilding	-4.33	0.00	-5.18	-8.33
25 Aerospace equipment	-1.79	-8.78	-12.54	0.53
26 Electrical engineering	4.54	-9.31	-7.06	-1.74
27 Instrument engineering	1.68	0.00	-6.13	-2.70
28 Engineers' small tools	3.18	-0.51	-6.56	-1.74
29 Music instruments, toys, etc	2.27	0.00	-4.87	-4.76
30 Timber processing	0.01	-3.59	-8.16	-6.30
31 Wooden furniture	9.35	-3.17	-0.74	-2.48
32 Pulp, paper, board	-0.39	0.54	-10.55	-1.86
33 Paper and board products	-0.99	0.00	-5.17	-1.50
34 Printing and publishing	4.40	-100.00	-5.55	-1.14
35 Leather, leather goods, footwear	18.72	-16.25	-3.03	-3.68
36 Textile goods	3.36	0.44	-8.55	-3.23
37 Clothes	6.95	-3.70	-1.45	-2.69
38 Food	7.55	-2.65	-6.73	-2.17
39 Drink	9.64	-10.47	-5.42	-1.61
40 Tobacco	0.04	15.48	-4.88	-3.13
41 Construction	5.92	-0.53	-2.42	-1.38
42 Trade wholesale & retail	8.16	-0.33	-3.11	-0.05
43 Traffic & transport services	3.66	0.00	-3.18	-0.16
44 Telecommunications	1.73	-9.28	-5.46	0.63
45 Banking, finance, insurance, etc	5.18	0.00	-4.48	0.38
46 Hotels, catering, etc	7.57	-1.21	-5.56	-0.54
47 Other services	0.48	-14.26	-6.16	2.17
Total	1.10	2.31	-1.19	0.85

Table 13.10. Actual changes to fuel use (i.e. C): Germany.

13.4.2 Change in Fuel Mix with Constant Energy Efficiency Constraint

	Sector (% change) UK	Gas	Coal	Oil	Energy
1	Agriculture	0.00	-12.37	-3.09	-2.45
2	Forestry & fishing	0.00	0.00	-9.61	-8.30
3	Electricity; fossil generation	18.20	-3.27	5.90	-0.22
4	Electricity; other generation	0.00	0.00	0.00	0.00
5	Electricity; distribution	0.00	0.00	0.00	0.00
6	Gas	63.94	-35.34	-25.88	-17.13
7	Water	14.10	-28.61	12.63	3.11
8	Coal extraction, coke ovens, etc	4.82	7.62	1.19	-2.18
9	Extraction of metalliferous ores	50.97	-100.00	6.48	2.30
10	Extraction of mineral oil and gas	0.00	0.00	0.00	0.00
11	Chemical products	18.92	-13.99	-6.05	0.45
12	Mineral oil processing	2.83	-6.82	-5.73	0.19
13	Processing of plastics	14.17	1.65	4.19	12.83
14	Rubber products	21.91	-7.49	2.28	0.10
15	Stone, clay, cement	23.24	-5.07	-7.30	-2.73
16	Glass, ceramic goods	8.50	-17.86	-12.33	-8.36
17	Iron and steel, steel products	12.46	-1.31	-6.34	-4.55
18	Non-ferrous metals	7.52	3.86	-7.39	-1.06
19	Foundries	9.79	-6.61	-2.03	14.19
20	Production of steel etc	3.58	-10.53	-3.56	-3.96
21	Mechanical engineering	5.55	-13.84	-3.24	-2.65
22	Office machines	-3.64	-100.00	-12.79	-0.48
23	Motor vehicles	5.95	-3.32	-6.17	-2.67
24	Shipbuilding	1.43	-23.77	-5.05	-5.26
25	Aerospace equipment	8.07	-13.34	-10.20	-5.06
26	Electrical engineering	3.24	-18.44	-8.73	-3.67
27	Instrument engineering	4.06	-100.00	-7.24	-2.88
28	Engineers' small tools	5.54	-100.00	-2.58	-0.59
29	Music instruments, toys, etc	5.88	-28.67	1.01	-1.13
30	Timber processing	8.47	-13.30	10.61	6.11
31	Wooden furniture	14.25	-6.53	8.30	9.29
32	Pulp, paper, board	44.65	-8.30	-8.22	-5.91
33	Paper and board products	18.34	-14.00	-0.29	0.86
34	Printing and publishing	6.96	-19.68	-2.70	1.03
35	Leather, leather goods, footwear	14.90	-12.39	-8.08	-6.36
36	Textile goods	16.82	-9.26	-7.43	-7.15
37	Clothes	6.44	-17.88	-4.80	-2.76
38	Food	10.93	-10.56	-4.26	-0.73
39	Drink	22.54	-13.03	-5.99	-0.80
40	Tobacco	-100.00	-100.00	-100.00	-100.00
41	Construction	0.18	-100.00	-1.18	-2.43
42	Trade wholesale & retail	0.00	-100.00	-0.69	1.73
43	Traffic & transport services	0.00	-100.00	5.97	5.35
44	Telecommunications	0.00	-100.00	-5.97	-5.31
45	Banking, finance, insurance, etc	0.00	-100.00	-22.61	-3.31
46	Hotels, catering, etc	0.00	-7.94	-25.81	-7.38
47	Other services	0.00	-100.00	-17.88	-12.70
	Total	11.62	-5.21	-1.41	-2.50

Table 13.11. Actual changes to fuel use (i.e. C): UK.

Overall, the picture regarding improvements in fuel efficiency and inter-fuel substitution is very encouraging. These minimum disruption findings are precisely in line with the findings in Chapter 11.

13.5 Changes in the Structure of Inter-Industry Trading

The final type of minimum disruption restructuring we consider is that of the coefficients of the inter-industry trading matrix (i.e. the **A** matrix). In particular, for the sake of brevity we consider alterations in the row and column sums of the **A** matrix, necessary to reduce CO_2 emissions by 1% p.a., *ceteris paribus*. Here we apply the theory developed in Section 12.5.

We recall the interpretation of these changes. The changes to the column sums represent the change in the use of inputs to sectors; i.e. the improvement in efficiency of sectors in transforming produced inputs into output. For example, one might consider the improvement of input efficiency of the Clothes industry.

Changes to the rows represent the change in the use of the output of each sector, as inputs to other sectors. So here, the improvement in efficiency is related to a particular input commodity. For example, one might consider the improvement in efficiency of the use of Office Machines by all other sectors.

We seek an overall 1% p.a. reduction of CO_2 emissions by alterations to the **A** matrix, with 0.5% p.a. required from column alterations, and 0.5% p.a from row alterations. The results of this calculation are shown in Table 13.12.

Looking first at the required changes to the column sums of **A**, we see similar patterns for Germany and the UK. For Germany, the sectors where the required reductions exceed 1% p.a. are 5, 11, 21, 22, 25, 37, 41, 45, 47. For the UK, these sectors are: 5, 11, 12, 20, 26, 38, 41, 42, 43, 45. The interpretation of this finding is that these are the sectors which need to seek the greatest improvements in their input efficiencies. None of the required improvements in input efficiencies seem to be outrageous.

Examining the required changes to the row sums of the **A** matrix, we again see similar patterns for Germany and the UK. For Germany the only reductions greater than 1% p.a. are for sectors 3, 5 and 8, while for the UK these sectors are only 3 and 5.

As Sectors 3 and 5 relate to electricity production and distribution, the major interpretation of this finding is that it is the use of electricity *by other sectors* which it is important to improve in efficiency, in both countries.

13.5 Changes in the Structure of Inter-Industry Trading

Sector (% change)	Germany Column	Germany Row	UK Column	UK Row
1 Agriculture	-0.63	-0.11	-0.41	-0.12
2 Forestry & fishing	-0.06	-0.05	-0.07	-0.02
3 Electricity; fossil generation	-0.51	-2.42	-0.94	-3.40
4 Electricity; other generation	-0.17	-0.09	-0.08	-0.05
5 Electricity; distribution	-1.40	-2.62	-1.99	-3.27
6 Gas	-0.10	-0.01	-0.43	-0.03
7 Water	-0.05	-0.02	-0.07	-0.05
8 Coal extraction, coke ovens, etc	-0.31	-1.44	-0.31	-0.31
9 Extraction of metalliferous ores	-0.07	-0.09	-0.02	-0.04
10 Extraction of mineral oil and gas	-0.02	-0.02	-0.44	-0.03
11 Chemical products	-1.93	-0.33	-1.58	-0.37
12 Mineral oil processing	-0.17	-0.03	-1.29	-0.21
13 Processing of plastics	-0.49	-0.07	-0.33	-0.14
14 Rubber products	-0.10	-0.02	-0.12	-0.04
15 Stone, clay, cement	-0.39	-0.12	-0.48	-0.17
16 Glass, ceramic goods	-0.16	-0.07	-0.16	-0.06
17 Iron and steel, steel products	-1.86	-0.49	-0.74	-0.84
18 Non-ferrous metals	-0.31	-0.12	-0.24	-0.10
19 Foundries	-0.14	-0.04	-0.28	-0.13
20 Production of steel etc	-0.27	-0.03	-0.76	-0.14
21 Mechanical engineering	-1.49	-0.06	-1.30	-0.16
22 Office machines	-0.22	0.00	-0.16	-0.01
23 Motor vehicles	-2.27	-0.04	-0.89	-0.04
24 Shipbuilding	-0.06	-0.01	-0.16	-0.03
25 Aerospace equipment	-0.08	-0.01	-0.32	-0.01
26 Electrical engineering	-1.25	-0.06	-1.16	-0.09
27 Instrument engineering	-0.17	0.00	-0.14	-0.01
28 Engineers' small tools	-0.46	-0.05	-0.04	0.00
29 Music instruments, toys, etc	-0.07	0.00	-0.14	-0.01
30 Timber processing	-0.12	-0.07	-0.22	-0.08
31 Wooden furniture	-0.37	-0.01	-0.21	-0.01
32 Pulp, paper, board	-0.20	-0.19	-0.14	-0.16
33 Paper and board products	-0.25	-0.05	-0.29	-0.10
34 Printing and publishing	-0.28	-0.03	-0.61	-0.06
35 Leather, leather goods, footwear	-0.06	-0.01	-0.11	-0.02
36 Textile goods	-0.34	-0.08	-0.44	-0.09
37 Clothes	-0.26	-0.01	-0.21	-0.01
38 Food	-2.14	-0.09	-3.07	-0.14
39 Drink	-0.29	-0.03	-0.60	-0.05
40 Tobacco	-0.06	0.00	-0.09	0.00
41 Construction	-2.00	-0.05	-3.98	-0.06
42 Trade wholesale & retail	-1.69	-0.15	-3.76	-0.19
43 Traffic & transport services	-0.97	-0.27	-1.37	-0.62
44 Telecommunications	-0.08	-0.01	-0.42	-0.04
45 Banking, finance, insurance, etc	-3.13	-0.03	-3.16	-0.21
46 Hotels, catering, etc	-0.66	-0.04	-0.68	0.00
47 Other services	-3.16	-0.19	-0.61	-0.01

Table 13.12. Minimum disruption changes to interindustry trading (i.e. **A**): 1% reduction in CO_2 emission by Germany and the UK.

13 'Minimum Disruption' Scenario Simulations

Sector (% change)	Germany Column	Germany Row	UK Column	UK Row
1 Agriculture	-1.71	-1.02	-1.19	1.93
2 Forestry & fishing	-0.41	-3.91	3.79	14.38
3 Electricity; fossil generation	3.49	-3.42	0.62	-0.70
4 Electricity; other generation	3.28	10.48	2.50	3.75
5 Electricity; distribution	0.00	0.70	0.04	4.27
6 Gas	-1.22	-0.61	-15.79	8.21
7 Water	5.58	-1.89	29.90	0.20
8 Coal extraction, coke ovens, etc	1.41	0.34	10.57	-16.82
9 Extraction of metalliferous ores	-0.15	1.39	8.19	-8.60
10 Extraction of mineral oil and gas	-1.93	-0.38	0.00	0.00
11 Chemical products	-0.49	-1.47	-2.59	-3.68
12 Mineral oil processing	2.29	-4.30	4.87	-2.18
13 Processing of plastics	0.02	0.92	-2.88	9.48
14 Rubber products	0.85	-1.34	-2.31	-6.20
15 Stone, clay, cement	1.10	-1.50	-0.26	0.98
16 Glass, ceramic goods	1.05	0.52	0.41	-5.25
17 Iron and steel, steel products	-0.67	-1.43	-1.01	-4.22
18 Non-ferrous metals	1.06	-1.95	-2.91	-7.29
19 Foundries	1.11	-2.45	-3.37	16.29
20 Production of steel etc	0.96	-1.67	0.39	-6.44
21 Mechanical engineering	0.77	-2.06	0.73	-0.45
22 Office machines	0.03	-5.45	-5.35	28.04
23 Motor vehicles	1.70	-1.80	-4.66	-5.85
24 Shipbuilding	2.69	2.71	-1.66	-0.69
25 Aerospace equipment	-0.65	3.73	-5.35	-15.99
26 Electrical engineering	-0.17	0.95	-4.08	-2.29
27 Instrument engineering	1.00	-2.38	-2.74	-7.25
28 Engineers' small tools	0.65	-0.17	-0.48	-18.15
29 Music instruments, toys, etc	3.15	-1.32	-2.54	-4.12
30 Timber processing	-1.34	-0.36	3.97	-1.59
31 Wooden furniture	1.70	-5.53	-0.64	-4.31
32 Pulp, paper, board	-0.05	-0.08	0.67	-3.75
33 Paper and board products	0.56	-1.43	-0.83	-0.45
34 Printing and publishing	1.47	-2.81	0.36	2.23
35 Leather, leather goods, footwear	0.73	-6.78	-2.81	-2.80
36 Textile goods	0.70	-0.78	-4.66	-4.47
37 Clothes	1.78	-3.81	-5.08	0.09
38 Food	-0.01	-3.40	0.96	5.81
39 Drink	1.61	-3.40	-1.26	7.61
40 Tobacco	4.73	-6.92	1.03	-13.40
41 Construction	0.56	3.26	2.17	-6.44
42 Trade wholesale & retail	-0.47	0.40	6.09	5.20
43 Traffic & transport services	0.57	1.99	5.60	-0.40
44 Telecommunications	-2.09	4.98	3.06	3.56
45 Banking, finance, insurance, etc	0.74	2.00	3.22	7.15
46 Hotels, catering, etc	-0.71	0.55	4.84	-43.17
47 Other services	-1.41	9.45	2.35	-16.28

Table 13.13. Actual annualised rates of change of the row and column sums of the A matrices: Germany (1978-88) and the UK (1968-84).

A word of warning is required here. The simplest way to reduce the elements of the A matrix is to shift from the use of domestically produced intermediate goods, towards imported ones. Recalling the discussion in Sections 8.4 and 10.3, this would, of course, reduce the *attributable* CO_2 emissions, but not the emissions for which that country is *responsible*. In any policy analyses concerning CO_2 emissions reduction, the 'attributable/responsible' distinction must always be borne in mind. The important figure is the CO_2 emissions for which a country is *responsible*.

In Table 13.13 is shown the actual changes in the row and column sums of the A for Germany (1978-88) and the UK (1968-84). Looking first at the data for Germany, we see that there are 16 negative column-sum changes, and 31 negative row-sum changes. In particular, the changes in the row-sums are often considerably negative, indicating greatly improved input efficiency of the outputs of these sectors. However, in both column- and row-sums there are some several large positive elements.

Looking now at the data for the UK we see there are 23 negative column-sum changes, and 29 row-sum changes. This is very encouraging, especially regarding the row-sum elements. However, the row-sum change for Electricity Distribution (5) is 4.27, indicating that the use of electricity in production is becoming more important over time. In view of the high CO_2 intensity of electricity, this is a disappointing finding if the aim is to reduce overall CO_2 emissions.

Overall, the mainly negative character of these changes is quite encouraging, and in line with the finding discussed in Section 10.4.2. There we noted that the contribution of changes in the Leontief inverse to changes in CO_2 emission was mainly negative, for both Germany and the UK.

13.6 Conclusions

In this chapter we have applied the minimum disruption approach to scenario analysis, which we derived in Chapter 12. We used a target reduction of CO_2 emission by production of 1% p.a., together with growth and employment targets, and an energy efficiency constraint. We explored scenarios for this reduction in CO_2 emissions by considering changes in final demand (**y**), energy use (**C**), and inter-industry trading (**A**). We compared the calculated minimum disruption scenario rates of change with those actually observed from our data set. Overall, the results are encouraging.

For changes in the final demand vector (**y**), we found that the minimum disruption changes, consistent with 2% p.a. growth and constant (effective) employment, are not extravagant. Regarding changes in fuel use mix and efficiency, the minimum disruption changes required are actually modest compared with the changes actually achieved, particularly for the UK. Inter-industry trading patterns are also evolving in a way largely consistent with reductions in CO_2 emissions.

In conclusion, the minimum disruption scenarios we have explored, which together would give rise to 3% p.a. reductions in CO_2 emissions from production, are largely consistent with observed trends, and also consistent with the 'best practice' improvements discussed in Chapter 12. Our conclusion is that reductions of 1% p.a. in CO_2 emissions by the production part of the German and UK economies are eminently achievable.

Part VI

Policy

14 Policy Conclusions for Reducing CO_2 Emissions

14.1 Introduction

In this final chapter we summarise our major conclusions in Section 14.2. We then offer, in Section 14.3, a brief commentary on how these conclusions can inform policy regarding reducing CO_2 emission rates by Germany and the UK, and the EC as a whole. Finally, in Section 14.4 we stress again the need for the *will* in confronting the global greenhouse effect.

At a 1988 conference on Global Warming in Toronto, a target was suggested for developed countries to reduce CO_2 emissions by 20% of 1988 levels by 2005. We approximate this target as a sustained 1% p.a. reduction in CO_2 emission rates, over 20 years. It is to this modified 'Toronto target' that our conclusions apply.

14.2 Major Conclusions

In this study we have undertaken a mixture of methodological (Chapters 1, 2, 3, 6), theoretical (Chapters 4, 7, 8, 12) and empirical work (Chapters 5, 9, 10, 11, 13). It should be noted that the two former categories were presented largely to support and motivate the final, empirical, category.

Our aim in this study was to explore how far structural economic change could be undertaken to reduce CO_2 emissions. In particular, we have studied in considerable detail the historical structural change of the German and UK economies, and how this has been reflected in changing CO_2 emissions. This has allowed us to construct scenarios for reducing CO_2 emissions for both countries.

Generally, our findings have been very encouraging for those wishing to meet our Toronto target. We now detail the major conclusions from our study which have long-run policy implications. We draw some overall conclusions in the final section. We classify these into three categories: History, Analysis, and Scenarios.

14.2.1 History

1. There is evidence that increasing atmospheric CO_2 concentrations increase the risk of major global climate change (Chapters 1 and 3).

2. The rate of increase of World CO_2 emissions is presently about 2% p.a., which is down from 5% p.a. in the early sixties (Figure 4.1). Although this is an encouraging trend, it indicates the importance of adopting global policies to reduce, and reverse, this rate of increase of CO_2 emissions.

3. The World energy/output ratio is falling at about 2% p.a., while the CO_2/energy ratio is roughly constant (Figure 4.1). This is encouraging, indicating that global economic growth does not *necessarily* need to be sacrificed in the pursuit of reducing CO_2 emissions.

4. For the EC, CO_2 emissions have been falling at about 2% p.a. in recent years. This is *twice* the rate of reduction necessary to meet our modified Toronto target (Figure 4.3).

5. For the EC, in recent years the energy/output ratio has been falling at about 4% p.a., while the CO_2/energy ratio has been relatively constant. This has been consistent with sustained growth in GDP of about 2% p.a. (Figure 4.3).

14.2.2 Analysis

6. German CO_2 emissions have been falling at about 1.5% p.a. since the late seventies. This can be attributed to a falling energy/output ratio and a falling CO_2/energy ratio. At the same time, GDP growth of more than 2% p.a. has been achieved (Figure 5.22).

7. UK CO_2 emissions have also been falling, at about 1% p.a., since the late seventies. This can be attributed to falling energy/output and CO_2/energy ratios. This has allowed GDP growth of about 2% p.a. to be maintained (Figure 5.23).

8. CO_2 emissions attributable to fuel use in production have been falling in both countries. The rate of fall in Germany has 1-2% p.a. since the mid-seventies, with the fall in the UK since the early seventies being at the rate 1-3% p.a. (Figures 5.24 and 5.25).

9. CO_2 emissions attributable directly to household fuel use have been approximately constant since the late seventies, in both Germany and the UK (Figures 5.26 and 5.27).

10. For the UK there is evidence that, while most sectors are becoming more efficient in their use of energy, there is evidence of significant *worsening* energy efficiency in the Transport sector (Figure 5.31).

11. Since 1981 in the UK, when the changing mix of producing sectors is taken into account, there has been an overall *worsening* in the efficiency of generating GDP from energy. Therefore, the observed apparent improvement in the efficiency of overall energy use in the UK can be largely attributed to the shift from industrial activity towards services (Figure 5.37).

12. The observed structural changes, changes in CO_2 emissions over time, and the results of our scenario analyses, are very similar for Germany and the UK. Therefore, we think that we could draw conclusions for other industrialised countries, especially for the EC, from our study (*passim*).

13. For both Germany and the UK, the proportion of CO_2 emissions attributable directly to the domestic final demand for fuels has been increasing, from about 20% in 1968 to about 30% in 1988. The proportion of CO_2 emissions attributable to the *direct* use of energy in production has been steady, at about 17% for both Germany and the UK. The proportion attributable to the *indirect* use of energy in production has fallen from about 60% in 1968 to about 55% in 1988, for both countries (Figure 9.1).

14. A large proportion of the direct *plus* indirect CO_2 emissions are attributable to Mineral Oil Processing; Electricity; Chemical Products. This pattern is true over time, and between countries (Section 9.5.5).

15. When imports and exports are properly accounted for, both Germany and the UK are 'responsible' for less CO_2 than they actually emit. The UK is 'responsible' for about 95% of its emissions, while Germany is 'responsible' for about 93% of its total CO_2 emissions (Figure 10.10).

16. For both countries, there is evidence that while there has been some improvement in the efficiency of energy use in production, there is less evidence for such efficiency improvements in the direct use of energy by households, government expenditure and investment (Figure 10.5).

274   14 Policy Conclusions for Reducing CO_2 Emissions

17. For both countries, there is evidence that the rise of the 'Service' industries has led to a fall in total CO_2 emissions (Figure 10.5).

18. The German mix of production is less CO_2 intensive than the UK mix (Table 10.5).

19. The 'minimum disruption' approach can be taken to alterations in economic structure, to achieve target CO_2 emission reduction rates. This allows us to identify those CO_2 reduction measures which give 'minimum disruption' to the economic structure, while still attaining target rates of (e.g.) GDP growth and employment growth (Chapter 12).

14.2.3 Scenarios

20. In the absence of economic growth, the introduction of best practice technologies of energy use in production would lead to CO_2 emissions falling by about 1.4% p.a. in both countries (Table 11.4). Best practice technologies in household energy use would lead to CO_2 emissions falling by about 0.6% p.a. in both countries (Table 11.5). Therefore the total rate of decrease of CO_2 emissions, with the introduction of best practice technology and in the absence of economic growth, would be 2% p.a.

21. The complete substitution of coal by gas in both countries would, in the absence of economic growth, lead to a 'once-off' reduction in total CO_2 emissions by about 11%. The substitution of gas for coal only in electricity generation leads to a 'once-off' reduction in CO_2 emissions by about 7-8% for both economies. These reductions of CO_2 emissions are, however, insufficient to meet our Toronto target (Section 11.4.1).*

22. Given an overall growth rate of GDP of 2% p.a., with a higher growth rate in the services (4% p.a.) and food (6% p.a.) sectors, and a lower rate of growth in the remaining sectors, an improvement in efficiency of energy use by industry and households of about 2% p.a. and a switch to renewable energy sources by about 1% p.a., are sufficient to meet our Toronto target. This is true *even* with the phasing out of nuclear power. (Table 11.11).

23. From the application of the 'minimum disruption' approach to scenario analysis, it seems that the necessary changes to meet our Toronto target for CO_2 emission reduction, for the production side of the economy, are close to those already being achieved. This is particularly true for fuel substitution (Tables 13.8-13.10). Also, the 'minimum disruption' changes in the structure of final demand compatible with reaching the Toronto target, maintaining GDP growth at 2% p.a., and maintaining full employment are demanding, but not too dissimilar from the historical trends observed (Tables 13.4, 13.5).

14.3 Policy Overview

What do these major conclusions tell us? First, there is a global problem of increasing CO_2 emissions, which needs to be confronted. Second, the developed countries, while principally responsible for these CO_2 emissions, are also in a position to exploit their on-going structural alterations to reduce their CO_2 emissions substantially.

In particular, there is a fortunate three-fold tendency available for further advancement by policy makers. The three components are:

1. A shift away from coal, and towards less CO_2 intensive oil and, especially, gas.

2. Improvements in the efficiency of energy use, particularly for use in production, and to a lesser extent by households.

3. A shift from heavy manufacturing towards more service and 'information' based economic activity.

These already existing trends could be encouraged in a number of relatively easily implemented ways. For example, numerous studies have shown that 'carbon taxes' are likely to lead to changed consumer behaviour, and changed techniques of production, which would substantially reduce CO_2 emissions.

However, in the UK one sector is showing significant *deterioration* in energy efficiency; the transport sector. This can be attributed to the marked shift away from public transport in recent decades. However, improved subsidies to public transport, and 'road charging' in cities could reverse this trend.

Also, there is ample scope for improving the energy efficiency of households. Policies here could include:

a. Subsidies on improved home insulation.

b. Higher insulation standards required in new houses.

c. Differential taxes on energy using home appliances, with higher taxes on inefficient appliances.

Within the production side of the economy, carbon taxes would probably provide a great incentive to the installation of new, energy efficient capital equipment (which would also aid productivity, as a side-effect). However, small firms may find it difficult to access sufficient financial capital to take advantage of such new equipment. For such small firms, a system of grants or low interest loans, to allow re-equipping, could be effective.

We should introduce one caveat regarding policies to reduce CO_2 emissions. It is completely fruitless to pursue policies which merely 'export' our CO_2 emissions to other countries. For example, a high and unilateral carbon tax will have the effect of making it more profitable to shift 'CO_2 intensive' production to other (possibly less developed) countries. Such an outcome will be totally worthless regarding the reduction of *global* CO_2 emissions. The global problem of the Anthropogenic Greenhouse Effect requires a global solution.

In particular, it requires the rich, industrialised countries to accept their *joint* responsibility for the over use of this global atmospheric commons, and to formulate *joint* policies for reforming their polluting ways.

14.4 The Need for the *Will*

In Chapter 1 we began with an overview of the global warming problem, resulting from the Anthropogenic Greenhouse Effect. In that chapter we concluded that one of the most important aspects in dealing with the CO_2 problem was achieving the *will* that a solution *would* be achieved.

We close our study by reference again to the necessity of will. From our study we are convinced that the reduction of CO_2 emissions by Germany and the UK (and probably by the EC and the rest of the industrialised world) is relatively easily achieved. Current trends in technology and consumption are already guiding us in the right direction. We *can* achieve our Toronto target of sustained 1% p.a. reductions in CO_2 emissions, and *still* enjoy reasonable rates of economic growth and employment.

Further, we do not accept the arguments put forward by some economists that this target reduction in CO_2 emissions has a significant, even large, cost. Certainly, the economic structure we would seek to establish would be different from that we now have, or might expect for the future without policy intervention. One aspect in which it would be different is that the capital stock would be not only more energy efficient, but almost certainly more *economically* efficient also. Much technological progress has been by the invention of new techniques which are more efficient in *many* more ways than the old techniques they replace. For example, weight for weight, the modern jet engine is not only far more powerful, but far more fuel efficient, than the internal combustion engine it replaced.

So, we conclude that *if* there can be established the *will* to reduce CO_2 emissions, both within and between the industrialised countries, then the outcome is eminently achievable, without long-run economic cost. As the prospect of global warming is perhaps the greatest threat facing humankind, we sincerely hope that our political leaders can summon the foresight, wisdom and determination to implement the necessary policies.

References

Adelman, M.A. (1980) "Energy-income coefficients and ratios", Energy Economics 2:2-4.
Alfsen, K.H. (1992) "Use of macroeconomic models in analysis of environmental problems in Norway and consequences for environmental statistics", Statistical Journal of the United Nations, ECEG, 51-72.
Allen, R.I.G. (1976) "The energy coefficient and the energy ratio", Economic Trends No. 274:78-84.
Axelrod, R. (1984) The Evolution of Co-operation. Basic Books, New York.

Barbier, E.B. and D.W. Pearce (1990) "Thinking economically about climate change", Energy Policy 18:11-18.
Baumol, W.J. (1977) Economic Theory and Operations Analysis (4th edn.). Prentice/Hall International, London.
Bergman, L. (1991) "General equilibrium effects of environmental policy: a CGE-modeling approach", Environmental and Resource Economics 1:67-85.
Bernholz, P. (1971) "Superiority of roundabout processes and positive rate of interest: a simple model of capital and growth", Kyklos 24:687-721.
Bernholz, P. and M. Faber (1976) "Time consuming innovation and positive rate of interest", Zeitschrift für Nationalökonomie 36:347-367.
Berry, K. and S. Loudenslager (1987) "The impact of nuclear power plant construction activity on the electric utility industry's cost of capital", Energy Journal 8:63-75.
Bliss, C.J. (1975) Capital Theory and the Distribution of Income. North Holland, Amsterdam.
Blitzer, C.R., R.S. Eckhaus, S. Lahiri and A. Meeraus (1990a) A General Equilibrium Analysis of the Effects of Carbon Emission Restrictions on Economic Growth in a Developing Economy. MIT, Department of Economics, Working Paper # 558.
Blitzer, C.R., R.S. Eckhaus, S. Lahiri and A. Meeraus (1990b) The Potential for Reducing Carbon Emissions from Increased Efficiency: A General Equilibrium Methodology. MIT, Department of Economics, Working Paper # 559.
Boero, G., R. Clarke and L.A. Winters (1991) The Macroeconomic Consequences of Controlling Greenhouse Gases: A Survey. Department of the Environment, HMSO, London.
Bolin, B., B.R. Döös, J. Jäger, and R.A. Warrick (1986) The Greenhouse Effect, Climatic Change and Ecosystems. Wiley, New York.
Bölkow, L., M. Meliß and H.J. Ziesing (1990) Erneuerbare Energiequellen, Endbericht der Koordinatoren zum Studienschwerpunkt A2 aus: Studienprogramm für die Enquetekommission 'Vorsorge zum Schutz der Erdatmosphäre' des Deutschen Bundestages. Berlin u.a.
Brown, G.M. and R.W. Johnson (1982) "Pollution control by effluent charges: it works in the Federal Republic of Germany, why not in the USA?", Natural Resource Journal 22:929-966.
Burmeister, E. (1980) Capital Theory and Dynamics. Cambridge University Press, Cambridge.
Burmeister, E. and A.R. Dobell (1970) Mathematical Theories of Economic Growth. Macmillan, New York.

CBO (1990) Carbon Charges as a Response to Global Warming: the Effect of Taxing Fuels. Congressional Budget Office, Congress of the United States, Washington.

Central Statistical Office (1973) Input-output Tables for the United Kingdom 1968. HMSO, London.
Central Statistical Office (1980) Annual Abstract of Statistics 1979. HMSO, London.
Central Statistical Office (1980) Input-output Tables for the United Kingdom 1974. HMSO, London.
Central Statistical Office (1983) Input-output Tables for the United Kingdom 1979. HMSO, London.
Central Statistical Office (1984) Annual Abstract of Statistics 1983. HMSO, London.
Central Statistical Office (1984) Input-output Tables for the United Kingdom 1984. HMSO, London.
Central Statistical Office (1987) Annual Abstract of Statistics 1986. HMSO, London.
Central Statistical Office (1988) Input-output Tables for the United Kingdom 1984. HMSO, London.
Chapman, D. (1990) The eternity problem: nuclear power waste storage, Contemporary Policy Issues 8:80-93.
Cline, W.R. (1991) "Scientific basis for the greenhouse effect", Economic Journal 101:904-919.
Conrad, K. and I. Henseler-Unger (1986) "The economic impact of coal-fired versus nuclear power plants: an application of a general equilibrium model", Energy Journal 7:51-63.
Conrad, K. and M. Schröder (1991) "The control of CO_2 emissions and its economic impact: an AGE model for a German state". In: H. Siebert (ed.), Environmental Scarcity: The International Dimension, J.C.B. Mohr, Tübingen.
Costanza, R. (1990) "What is ecological economics?", Ecological Economics 1:1-7.
Costanza, R. (ed.) (1991) Ecological Economics: The Science and Management of Sustainability. Columbia University Press, New York.
Costanza, R., B. Norton and B. Haskell (eds.) (1992) Ecosystem Health: New Goals for Environmental Management. Island Press, Washington.
Cropper, M.L. (1980) "Pollution aspects of nuclear energy use", Journal of Environmental Economics and Management 7:334-52.

Daily G.C., P.R. Ehrlich, H.A. Mooney and A.H. Ehrlich (1991) "Greenhouse economics: learn before you leap", Ecological Economics 4:1-10.
Daly, H.E. (1991) "Elements of environmental macroeconomics". In: Costanza (ed.) (1991).
Daly, H.E. and J.B. Cobb (1990) For the Common Good. Greenprint, London.
Darmstadter, J. (1991) "Estimating the cost of carbon dioxide abatement", Resources 103:6-9.
Dasgupta, P.G. (1983) The Control of Resources. Harvard University Press, Cambridge, Mass.
Department of Energy (1981) Digest of United Kingdom Energy Statistics 1980. HMSO, London.
Department of Energy (1989) Digest of United Kingdom Energy Statistics 1988. HMSO, London.
Department of Industry (1975) Report of Census of Production 1974. HMSO, London.
Department of Industry (1980) Report of Census of Production and Purchases Enquiry 1979. HMSO, London.
Department of Trade and Industry (1987) Purchases Inquiry 1984. HMSO, London.
Department of Transport (1987) Transport Statistics Great Britain 1976-1986. HMSO, London.

DIW/ISI (1984) Erneuerbare Energiequellen. Abschätzung des Potentials in der Bundesrepublik Deutschland bis zum Jahre 2000. Berlin: Deutsches Institut für Wirtschaftsforschung, und Karlsruhe: Fraunhofer Institute for Systems and Innovation Research.

DIW/ISI (1991) Kostenaspekte erneuerbarer Energiequellen in der Bundesrepublik Deutschland und auf Exportmärkten. Untersuchung im Auftrag des Bundesministers für Forschung und Technologie (forthcoming).

Dixon, P.B., D.T. Johnson, R.E. Marks, P. McLennan, R. Schodde and P.L. Swan (1989) The Feasibility and Implication for Australia of the Adoption of the Toronto Proposal for Carbon Dioxide Emissions. Report to CRA, Sydney.

Dorfman, R., P.A. Samuelson and R. Solow (1958) Linear Programming and Economic Analysis. McGraw-Hill, New York.

Eissenbeiss, G., A. Pluger and R. Windheim (1989) Government Support for the Development of Wind Energy Utilisation in the Federal Republic of Germany. Proceedings of the European Wind Energy Conference, September 1989, Glasgow.

Energie-Statistisches Jahrbuch 1988 (1990) Eurostat, Luxemburg.

Energy Statistics Yearbook (1989) United Nations, Statistical Office, New York.

Faber, M. and R. Manstetten (1989) Rechtsstaat und Umweltschutz aus ökonomischer Sicht, Zeitschrift für angewandte Umweltpolitik 3:361-371.

Faber, M., R. Manstetten and J.L.R. Proops (1992a) "Towards an open future: ignorance, novelty and evolution". In: Costanza, Norton and Haskell (eds.) (1992).

Faber, M., R. Manstetten and J.L.R. Proops (1992b) Humankind and the environment: an anatomy of surprise and ignorance, Environmental Values (forthcoming).

Faber M., H. Niemes and G. Stephan (1983) Umweltschutz und Input-Output Analyse. Mit zwei Fallstudien aus der Wassergütewirtschaft (Environmental Protection and Input-Output Analysis. With Two Case Studies in Water-Quality Management). J.C.B. Mohr, Tübingen.

Faber, M., H. Niemes and G. Stephan (1987) Entropy, Environment and Resources: An Essay in Physico-Economics. Springer-Verlag, Heidelberg.

Faber, M. and J.L.R. Proops (1985) Interdisciplinary research between economists and physical scientist: retrospect and prospect, Kyklos 4:599-616.

Faber, M. and J.L.R. Proops (1990) Evolution, Time, Production and the Environment. Springer-Verlag, Heidelberg.

Faber, M. and J.L.R. Proops (1991a) "National accounting, time and the environment: a neo-Austrian approach". In: Costanza (ed.) (1991).

Faber, M. and J.L.R. Proops (1991b) Resource Rents and National Accounting: a Capital Theoretic Approach. Discussion Paper No. 167, Department of Economics, University of Heidelberg.

Faber, M. and J.L.R. Proops (1991c) "The innovation of techniques and the time-horizon: a neo-austrian approach", Structural Change and Economic Dynamics 2:143-158.

Faber, M., J.L.R. Proops, M. Ruth and P. Michaelis (1990) "Economy-environment interactions in the long-run: a neo-Austrian approach", Ecological Economics 2:27-55.

Faber, M. and G. Stephan (1987) "Umweltschutz und Technologiewandel". In: R. Henn (ed.), Technologie, Wachstum und Beschäftigung, Festschrift für Lotar Späth. Springer-Verlag, Heidelberg.

Faber, M., G. Stephan and P. Michaelis (1989) Umdenken in der Abfallwirtschaft (2nd edn.). Springer-Verlag Heidelberg.

Faber, M. and G. Wagenhals (1988) "Towards a long-run balance beetween economics and environmental protection". In: W. Salomon and U. Förstner (eds.), Environmental Impact and Management of Time Tailings and Dredged Material, Springer-Verlag, Heidelberg.

Fry, G. R. (1986) "The economics of home solar water heating and the role of solar tax credits", Land Economics 62:134-44.

Gay, P.W. and J.L.R. Proops (1992) "Carbon dioxide production by the U.K. economy: an input-output assessment", Applied Energy (forthcoming).

Georgescu-Roegen, N. (1971) The Entropy Law and the Economic Process. Harvard University Press, Cambridge/Mass.

Graham-Bryce, I.J. (1989) "Optimisation of energy strategies for power generation in relations to global climate change". In: OECD (1989a) 2:159-171.

Guentenspenger, H. (1990) "Die Bedeutung der Altersstruktur des Pkw-Bestandes fuer CO_2-Reduktionsstrategien im Individualstrassenverkehr", Zeitschrift für Energiewirtschaft 1:47-56.

Hahn, R.W. (1989) "Economic prescriptions for environmental problems: how the patient followed doctor's orders in adjusting to global warming", Journal of Economic Perspectives 3:95-114.

Haraden, J. (1989) "CO_2: production rates for geothermal energy and fossil fuels", Energy 14:867-873.

Harding, J. (1990) "Reactor safety and risk issues", Contemporary Policy Issues 8:94-105.

Hasselmann K. (1991) "How well can we predict the climate crisis?". In: H. Siebert (ed.), Environmental Scarcity: the International Dimension, J.C.B. Mohr, Tübingen.

Hawkins, D. and H.A. Simon (1949) "Note: Some conditions of macroeconomic stability", Econometrica 17:245-248.

Hazilla, M., Kopp, R.J. (1990) "Social cost of environmental quality regulations: a general equilibrium analysis", Journal of Political Economy 98:853-873.

Heinze, F.G.R. (1989) The Economics of Nuclear Power Plants As They Age: Production (Capacity Factors), Variable Costs, and Retirement. Ph.D. Dissertation, Cornell University.

Hoeller, P., A. Dean and J. Nicolaisen (1991) "Macroeconomic implications of reducing greenhouse gas emissions: a survey of empirical studies", OECD Economic Studies No. 16:45-78.

Hohmeyer, O. (1988) Soziale Kosten des Energieverbrauchs. Springer-Verlag, Heidelberg.

IPCC (1990) Intergovernmental Panel on Climate Change, Climate Change: The IPCC Scientific Assessment. R. Houghton, G.J. Jenkins and E. Ephraumes (eds.), Cambridge University Press, Cambridge.

Johnston, J. (1972) Econometric Methods (2nd edn.). McGraw-Hill Kogakusha, Tokyo.

Khazzoom, J.D. (1987) "Energy saving resulting from the adoption of more efficient appliances", Energy Journal 8:85-89.

Kokoski, M.F. and V.K. Smith (1987) "General equilibrium analysis of partial-equilibrium welfare measures: the case of climate change", American Economic Review 77:331-341.

Kolb, G., G. Eickhoff, M. Kleemann, N. Krzikalla, M. Pohlmann, and H.J. Wagner (1989) CO_2 Reduction Potential Through Rational Energy Utilization and Use of Renewable Energy Sources in the Federal Republic of Germany. Jül-Spez-502, Kernforschungsanlage Jülich, Programmgruppe Systemforschung und Technologische Entwicklung.

Koopmanns, T.C. (1951) "Analysis of production as an efficient combination of activities". In: T.C. Koopmanns (ed.), Activity Analysis of Production and the Allocation of Resources, Wiley, New York.

Lee, H. and J.-C. Ryu (1991) "Energy and CO_2 emissions in Korea: long-term scenarios and related policies", Energy Policy 19:926-933.

Leontief, W. (1936) "Quantitative input-output relations in the economic system of the United States", Review of Economics and Statistics 18:105-125.

Leontief, W. (1951) The Structure of the American Economy, 1919-1939. Oxford University Press, London.

Leontief, W. (1966) Input-Output Economics, Oxford University Press.

Leontief, W. (1971) Theoretical assumptions and non-observed facts, American Economic Review 61:1-7.

Leontief, W. (1973) "National income, economic structure, and environmental externalities". In: M. Moss (ed.), Studies in Income and Wealth, Vol. 38, Columbia University Press, New York.

Maier, G. (1984) Rohstoffe und Innovationen. Eine dynamische Untersuchung (Resources and Innovation. A Dynamic Investigation). Mathematical Systems in Economics 68, Athenäum, Hain, Scriptor, Hanstein, Königstein/Ts.

Maier, W. and G. Angerer (1986) Rationelle Energieverwendung durch neue Technologien. 2 volumes, Köln.

Malinvaud, E. (1985) Lectures on Microeconomic Theory (2nd edn.). North Holland, Amsterdam.

Manne, A.S. (1990) GLOBAL 2100: An Almost Consistent Model of CO_2 Emission Limits. Paper presented at the Applied Equilibrium Modeling Conference, University of Bern, mimeo.

Manne, A.S. and R.G. Richels (1990a) Global CO_2 Emission Reduction: The Impacts of Rising Energy Costs. Electric Power Research Institute, Palo Alto, CA.

Manne, A.S. and R.G. Richels (1990b) "International trade in carbon emission rights: a decomposition procedure", American Economic Review 81:135-139.

Manne, A.S. and Richels, R.G. (1991) "Global CO_2 Emission Reductions. The Impact of Rising Energy Costs", Energy Journal 12:87-107.

Michaelis, P. (1991a) Theorie und Politik der Abfallwirtschaft. Eine ökonomische Analyse (Theory and Policy of Waste Management: An Economic Analysis). Studies in Contemporary Economics, Springer-Verlag, Heidelberg.

Michaelis, P. (1991b) Effiziente Klimapolitik im Mehrschadstoffall. Kieler Arbeitspapiere Nr. 456.

Miller, R.E. and P.D. Blair (1985) Input-Output Analysis: Foundations and Extensions. Prentice-Hall, Englewood Cliffs, New Jersey.

Morgenstern, R.D. (1991) "Towards a comprehensive approach to global climate change mitigation", American Economic Review 81:140-145.

Musgrove, P.J. (1990) "Electricity privatisation and its implications for wind energy". In: T.D. Davies, J.A. Halliday and J.P. Palutikov (eds.), Proceedings of the 12th British Wind Energy Association Conference, Norwich.

Nitschke, J. (1988) "Stromerzeugungskosten aus Windkraftanlagen", Elektrizitätswirtschaft 87:835-840.
Nordhaus, W.D. (1991) "To slow or not to slow: the economics of the greenhouse effect", Economic Journal 101:920-937.
Norgaard, R.B. (1989) "Three dilemmas of environmental accounting", Ecological Economics 1:303-314.

OECD (1985) Main Economic Indicators. Dept. of Economics and Statistics.
OECD (1988) Environmental Implications of Energy Use in Industry. Paris.
OECD (1989) Energy Statistics 1970-85. Paris.
OECD (1989a) Energy Technologies for Reducing Emissions of Greenhouse Gases. Proceedings of an Experts' Seminar, Paris, 12th-14th April 1989, 2 volumes, Paris.

Page, D.I., L.A.W. Bedford, P.L. Surman, D.J. Milborrow and W.G. Stevenson (1989) Large Scale Wind Energy Systems in the UK. Proceedings of the European Wind Energy Conference, September 1989, Glasgow.
Palz, W., J. Coombs and D.O. Hall (eds.) (1985) Energy from Biomass. Elsevier, London.
Pearce, D.W. (1991a) Personal communication.
Pearce, D.W. (1991b) "The role of carbon taxes in adjusting to global warming", Economic Journal 101:938-948.
Pearson, P.J.G. (1986) Input-Output Analysis and Air Pollution. Surrey Energy Economics Discussion Paper No. 30.
Pearson, P.J.G. (1989) "Proactive energy-environment policy strategies: a role for input-output analysis?", Environment and Planning A 21:1329-1348.
Petersen, Th. (1992) "James M. Buchanans 'The Limits of Liberty' und G.W.F. Hegels 'Rechtsphilosophie'". Heidelberger Diskussionspapier Nr. 174.
Proops, J.L.R. (1977) "Input-output analysis and energy intensities: a comparison of some methodologies", Applied Mathematical Modelling 1:181-186.
Proops, J.L.R. (1984) "Modelling the energy-output ratio", Energy Economics 6:47-51.
Proops, J.L.R. (1988) "Energy intensities, input-output analysis and economic development". In: M. Ciaschini (ed), Input-output Analysis: Current Developments, Chapman and Hall, London.
Proops, J.L.R. (1990) "Ecological economics: rationale and problem areas", Ecological Economics 1:59-76.

Quesnay, F. (1758) Tableau Economique.

Reilly, J.M., J.A. Edmonds, R.H. Gardner and A.L. Brenkert (1987) "Uncertainty analysis of the IEA/ORAU CO_2 emissions model", Energy Journal 8:1-29.
Rotty, R.M. and C.D. Masters (1985) "Carbon dioxide from fossil fuel combustion: trends, resources, and technological implications". In: Atmospheric Carbon Dioxide and the Global Carbon Cycle, United States Department of Energy.

Schäfer, D. and C. Stahmer (1989) Input-Output-Modelle zur gesamtwirtschaftlichen Analyse von Umweltschutzaktivitäten, ZfU 2, 127-158.
Schmitt, D. and H. Junk (1984) "The comparative costs of nuclear and coal-fired power stations in West Germany". In: L.G. Brookes and H. Motamen (eds.), The Economics of Nuclear Energy, Chapman and Hall, London.

Schmutzler, A. (1991) Flexibility and Adjustment in Sequential Decision Problems: A Systematic Approach. Lecture Notes in Economics and Mathematical Systems 371, Springer-Verlag, Heidelberg.

Simon, H.A. (1972) "Theories of bounded rationality". In: C.B. McGuire and R. Radner (eds.), Decision and Organisation, North-Holland, Amsterdam.

Sraffa, P. (1960) The Production of Commodities by Means of Commodities. Cambridge University Press, Cambridge.

Statistisches Bundesamt (1989) Volkswirtschaftliche Gesamtrechnungen: Fachserie 18. Metzler Poeschel, Stuttgart.

Statistisches Bundesamt (1991) Statistisches Jahrbuch. Metzler-Poeschel, Stuttgart.

Stephan, G., R. van Nieuwkoop and T. Wiedmer (1991) Kohlendioxid-Abgabe und wirtschaftliche Auswirkungen. Eine berechenbare Gleichgewichtsanalyse für die Schweiz. Diskussionsbeiträge des Volkswirtschaftlichen Instituts der Universität Bern, 91-3.

Stephan, G. (1989) Pollution Control, Economic Adjustment and Long-Run Equilibrium: A Computable Equilibrium Approach to Environmental Economics. Springer-Verlag, Heidelberg.

Streb, A. (1989) "Energy efficiency and global warming". In: OECD (1990) vol. 1.

Symons, E., J.L.R. Proops and P.W. Gay (1991) Carbon Taxes, Consumer Demand and CO_2 Emission: A Simulation Analysis for the UK. Working Paper in Economics and Management Science, University of Keele.

Umweltbundesamt (1988) Stellungnahme zum Fragenkatalog der Enquete-Kommission vom 25.4.1988 zum Thema 'Treibhauseffekt'. Umweltbundesamt, Berlin.

United Nations (1976) World Energy Supplies 1950-1976. Statistical Papers Series J, No. 19.

United Nations (1986) Yearbook of Industrial Statistics, 1986, Vol. 1.

United Nations (1989) National Accounts Statistics: Main Aggregates and Detailed Tables, 1986, part 1.

Victor, P.A. (1972) Pollution: Economy and Environment. Allen and Unwin, London.

Wagner, H.J. and G. Kolb (1989) "CO_2 emissions for Germany: present and projected trend with a technology scenario for their reduction". In: OECD (1989a) vol 2, 345-354.

Walton, A.L. and D.C. Hall (1990) "Solar power", Contemporary Policy Issues 8:240-54.

West, R.E. and F. Kreith (eds.) (1988) Economic Analysis of Solar Thermal Energy Systems, Solar Heat Technologies: Fundamentals and Applications, vol. 3. MIT Press, Cambridge/Mass.

Whalley, J. and R. Wigle (1990) The International Incidence of Carbon Taxes. Paper presented at conference on Economic Policy Responses to Global Warming, Rome, 4-6 October 1990. National Bureau of Economic Research, Cambridge, MA, USA, and Wilfrid Laurier University, Waterloo, Canada.

Whalley, J. and R. Wigle (1991) "Cutting CO_2 emissions: the effects of alternative policy approaches", Energy Journal 12:109-24.

Wodopia, F.J. (1986a) Intertemporale Produktionsentscheidungen: Theorie und Anwendung auf die Elektrizitätswirtschaft (Intertemporal Production Decisions: Theory and Application to the Electricity Sector). Mathematical Systems in Economics 102, Athenäum, Hain, Hanstein, Königstein/Ts.

Wodopia, F.J. (1986b) "Time and production: period versus continuous analysis". In: M. Faber (ed.), Studies in Austrian Capital Theory, Investment and Time, Springer-Verlag, Heidelberg.

Wodopia, F.J. (1986c) "Flow and fund approaches to irreversible investment decisions". In: M. Faber (ed.), Studies in Austrian Capital Theory, Investment and Time, Springer-Verlag, Heidelberg.

Zoutendjik, G. (1976) Mathematical Programming Methods. North-Holland, Amsterdam.

Author Index

Adelman, M.A., 67, 278
Alfsen, K.H., 97, 278
Allen, R.I.G., 67, 278
Angerer, G., 210, 282
Axelrod, R., 13, 278

Barbier, E.B., 40, 278
Baumol, W.J., 100, 231, 278
Bedford, L.A.W., 221, 283
Bergman, L., 97, 278
Bernholz, P., 23, 94, 278
Berry, K., 219, 278
Blair, P.D., 96, 99, 113, 121, 282
Bliss, C.J., 97, 278
Blitzer, C.R., 98, 210, 278
Boero, G., 45, 278
Bolin, B., 34, 35, 278
Bölkow, L., 218, 278
Brenkert, A.L., 210, 283
Brookes, L.G., 283
Brown, G.M., 11, 278
Burmeister, E., 25, 94, 97, 278

CBO, 44, 278
Chapman, D., 219, 279
Ciaschini, M., 283
Clarke, R., 45, 278
Cline, W.R., 19, 33, 34, 279
Cobb, J.B., 41, 279
Conrad, K., 97, 218, 279
Coombs, J., 222, 283
Costanza, R., 9, 10, 22, 279, 280
Cropper, M.L., 220, 279
CSO, 153, 154, 248, 279

Daily, G.C., 42, 279
Daly, H.E., 10, 41, 279
Darmstadter, J., 41, 279
Dasgupta, P.G., 9, 279
Davies, T.D., 282
Dean, A., 42, 45, 210, 281
Department of Energy, 65, 153-155, 279
Department of Industry, 153, 279
Department of Trade and Industry, 153, 154, 279
DIW/ISI, 218, 280
Dixon, P.B., 44, 280

Dobell, A.R., 25, 94, 278
Döös, B.R., 34, 35, 278
Dorfman, R., 92, 280

Eckhaus, R.S., 98, 210, 278
Edmonds, J.A., 210, 283
Ehrlich, A.H., 42, 279
Ehrlich, P.R., 42, 279
Eickhoff, G., 210, 215, 282
Eissenbeiss, G., 220, 280

Faber, M., 8, 11, 13, 15, 16, 20, 22, 23, 25-27, 41, 92, 94, 96, 97, 213, 278, 280, 281, 285
Fry, G.R., 218, 281

Gardner, R.H., 210, 283
Gay, P.W., 26, 46, 281, 284
Georgescu-Roegen, N., 7, 93, 281
Graham-Bryce, I.J., 211, 281
Guetensprenger, H., 209, 281

Hahn, R.W., 15, 281
Hall, D.C., 218, 284
Hall, D.O., 222, 283
Halliday, J.A., 282
Haraden, J., 221, 281
Harding, J., 219, 281
Haskell, B., 9, 279, 280
Hasselmann, K., 34, 281
Hawkins, D., 113, 281
Hazilla, M., 97, 281
Heinze, F.G.R., 219, 281
Henn, R., 280
Henseler-Unger, I., 218, 279
Hoeller, P., 42, 45, 210, 281
Hohmeyer, O., 218, 281

IPCC, 19, 33, 34, 281

Jäger, J., 34, 35, 278
Johnson, D.T., 44, 280
Johnson, R.W., 11, 278
Johnston, J., 134, 136, 281
Junk, H., 218, 283

Khazzoom, J.D., 215, 281

Author Index

Kleeman, M., 210, 215, 282
Kokoski, M.F., 97, 281
Kolb, G., 210, 211, 215, 282, 284
Koopmans, T.C., 93, 94, 102, 282
Kopp, R.J., 97, 281
Kreith, F., 218, 284
Krzikalla, N., 210, 215, 282

Lahiri, S., 98, 210, 278
Lee, H., 215, 282
Leontief, W., 91, 93, 96, 97, 99, 282
Loudenslager, S., 219, 278

Maier, A., 210, 282
Maier, G., 94, 282
Malinvaud, E., 92, 282
Manne, A.S., 42, 44, 45, 97, 282
Manstetten, R., 8, 13, 20, 280
Masters, C.D., 35, 40, 283
McGuire, C.B., 284
McLennan, P., 44, 280
Meeraus, A., 98, 210, 278
Meliß, M., 218, 278
Michaelis, P., 11, 13, 15, 16, 22, 26, 280, 282
Milborrow, D.J., 221, 283
Miller, R.E., 96, 99, 113, 121, 282
Mooney, H.A., 42, 279
Morgenstern, R.D., 42, 221
Moss, M., 282
Motamen, H., 283
Musgrove, P.J., 221, 282

Nicolaisen, J., 42, 45, 210, 281
Niemes, H., 22, 23, 26, 96, 280
Nieuwkoop, R. van, 97, 98, 284
Nitschke, J., 218, 283
Nordhaus, W.D., 42, 283
Norgaard, R.B., 41, 283
Norton, B., 9, 279, 280

OECD, 37-39, 210, 212, 281, 283, 284

Page, D.I., 221, 283
Palutikov, J.P., 282
Palz, W., 222, 283
Pearce, D.W., 6, 16, 278, 283
Pearson, P.J.G., 97, 121, 283
Petersen, Th., 13, 283
Pluger, A., 220, 280
Pohlmann, M., 210, 215, 282

Proops, J.L.R., 8, 13, 20, 22, 23, 25-27, 41, 46, 67, 92, 94, 96, 97, 280, 281, 283, 284

Quesnay, F., 99, 283

Radner, R., 284
Reilly, J.M., 210, 283
Richels, R.G., 42, 44, 45, 97, 282
Rotty, R.M., 35, 40, 283
Ruth, M., 26, 280
Ryu, J.-C., 215, 282

Samuelson, P.A., 92, 280
Schäfer, D., 96, 283
Schmitt, D., 218, 283
Schmutzler, A., 23, 92, 284
Schodde, R., 44, 280
Schröder, M., 97, 279
Simon, H.A., 24, 113, 281, 284
Smith, V.K., 97, 281
Solow, R., 92, 280
Sraffa, P., 100, 284
Stahmer, C., 96, 283
Statistisches Bundesamt, 152-154, 248, 284
Stephan, G., 11, 13, 15, 16, 22, 23, 26, 96-98, 280, 284
Stevenson, W.G., 221, 283
Streb, A., 209, 284
Surman, P.L., 221, 283
Swan, P.L., 44, 280
Symons, E., 26, 46, 284

Umweltbundesamt, 214, 284
UN, 37-39, 284

Victor, P.A., 44, 209, 284

Wagenhals, G., 26, 213, 281
Wagner, H.J., 210, 211, 215, 282, 284
Walton, A.L., 218, 284
Warrick, R.A., 34, 35, 278
West, R.E., 218, 284
Whalley, J., 97, 284
Wiedmer, T., 97, 98, 284
Wigle, R., 97, 284
Windheim, R., 220, 280
Winters, L.A., 45, 278
Wodopia, F.J., 26, 93, 284, 285

Ziesing, H.J., 218, 278
Zoutendjik, G., 229, 285

Subject Index

Activity analysis, 94 - 97
 and dynamics, 97
 and input-output analysis, 95
 and production functions, 92
Atmosphere
 Earth, 4

Boundary conditions
 and technology, 25
 natural, 23-25
Bounded rationality, 24

Capital goods
 accumulation, 7
Capital theory, 95, 97
 neoclassical, 22
Carbon convention, 16
Carbon tax, 16, 45
 correction of economic distortion, 16
 ecological efficiency, 17
 revenue consequences, 16
CFCs, 5, 11, 17, 21, 33
 global emissions, 35, 36
CO_2/energy ratio, 48, 49
 EC, 39, 67
 Germany, 67
 rate of change, 50, 54
 UK, 67
 USA, 38
 World, 38
CO_2/output ratio
 EC, 39, 68
 Germany, 68
 UK, 68
 USA, 39
 World, 39
CO_2 emissions
 20% reduction, 6
 and exports, 131, 174
 and final demand for fuels, 124
 and imports, 131, 132, 174
 and input-output analysis, 121
 and national income, 47
 attribution, 131, 133, 166
  to direct final demand, 171
  to export final demand, 171
  to final demand for fuels, 171
  to indirect final demand, 171
  Germany, 162, 170
  UK, 162, 170
  of differences, Germany, 181
  of differences, UK, 182
 by fuel type,
  Germany, 70
  UK, 70
 by sector, UK, 72, 73
 changes,
  energy efficiency effect, 85
  fuel mix effect, 85
  full sectoral decomposition, UK, 86
  output mix effect, 85, 86
  total output effect, 85
  attributable to final demand, 179
  attributable to fuel use change, 177, 178
  attributable to interindustry trading, 178
 criteria to assess reduction policies, 15
 decomposition of change, 47, 49, 74, 193
  by differences, 128, 129
  by sector, 82
  EC, 53
  Germany, 74, 75
  household, 77
   German, 78
   UK, 78
  production, 76
   German, 77
   UK, 77
  restructuring, 83
  total, 78
  UK, 74, 75
   Agriculture, 79
   Chemicals, 81
   Construction, 80
   Construction, 80
   Food, 81
   Other Ind., 82
   Paper etc, 81
   Services, 79
   Textiles, 81
   Transport, 80
  USA, 53

World, 52
derivative with respect to parameters, 140
difference between Germany and the UK, 188
difficulty of political implementation, 11
direct, 124
domestic, to meet domestic final demand, 137
EC, 37
effects of relative fuel prices, 130
elasticities, 136, 139, 144, 164, 227
  Germany 1988, 168
emission charges, 15
emission taxes, 15
foreign
  to meet domestic final demand, 138
  to meet imported final demand, 138
from non-fossil fuel sources, 125
  carbonaceous clays, 126
  glass making, 126
  incineration of waste materials, 126
fundamental input-output equation for, 149
Germany, 64, 65
  1988, 163
global, 34, 35
households, 71, 72
indirect, 124
input-output analysis,
  changes over time, 128
  data
    German, 149
    UK, 149
  Germany, 170
  Germany 1988, 157
  intensities, 123, 149
  inter-country, 128
  representation of total, 127
  UK, 170
multi-region analysis, 133
need for will, 276
practical implementation of reduction, 15
product of several variables, 48
production, 71, 72
ranked sectors, Germany 1988, 166
reduction of, 13
  conclusions, 271
    analysis, 272
    historical, 271
    scenarios, 274
  policy, 275
  targets, 6, 7, 11, 44
regional shares, 39
regulative laws, 15
responsibility, 136, 174, 175
  of developed countries, 14
sectoral,
  Germany 1988, 164
  decomposition of changes, 79
  UK, 83, 86
sensitivity to parameter changes, 138
stabilisation, 6
time structure, 47
tradeable emission permits, 15
UK, 64, 65
USA, 37
World, 37
CO_2 intensities, 158
  Germany 1988, 160
  ranked sectors, Germany 1988, 165
  sectoral, Germany 1988, 164
CO_2 responsibility
  direct imports, 176
  indirect imports, 177
  ratio to CO_2 emission, 177
  total imports, 177
Coal-miners' strike, 71
Consensus
  need for, 13, 17
Consumption
  impact on natural environment, 29
Conversion units, 36
Cost-benefit analysis, 10, 22
Cost functions, 42
Cropping pattern shifts, 34

Damage functions, 42
Data collection, 150
Data sources
  Germany, 65
  UK, 65
Difference
  approximation to differential, 50
  backward, 57, 58
  central, 57, 58
  decomposition remainder term, 54, 59, 75
  decomposition with, 55
  eight-period, 51

Subject Index

error-prone data, 61
forward, 57, 58
Differential
  approach to decomposition, 51, 55
  approximation with difference, 57
  discrete-time approximation, 50
Disruption
  CO_2 emission reduction target, 229, 230
    and employment growth target, 235
    and GDP growth target, 232
    GDP & and employment growth target, 238
  of final demand, 229, 248
    with employment constraint, 252
    with GDP & employment constraint, 252
    with GDP constraint, 250
  of fuel mix
    and constant energy efficiency, 241
    and fuel efficiency, 255
    efficiency constraint, 257
    fuel use coefficients, 240
  function, 228
  of inter-industry trading, 244, 264
  minimum, scenarios, 247
Dual representation, 100

Ecological Plimsoll lines, 10
Economics
  ecological, 10, 21, 22, 25
    modelling, 26
  environmental, 9
  neoclassical, 22
Ecosystem health, 9
Elasticities
  output/employment, 247
Electricity supply sector
  disaggregation, 151
Emission charges, 15
Emission taxes, 15
Energy/output ratio, 48, 49
  EC, 39, 68
  Germany, 68
  rate of change, 50, 54
  UK, 68
  USA, 39
  World, 39
Energy efficiency, 42
Energy intensity, 43
Energy use
  and GDP, 67
  by fuel type
    Germany, 70
    UK, 70
  EC, 37
  Germany, 64, 65
  household, 71, 72
  production, 71, 72
  UK, 64, 65
  USA, 37
  World, 37
Environmental economics, 9
Environmental institutions, 24
Environmental policy, 24
  need for economic efficiency, 15
  need for experience with instruments, 16
  need for flexibility of instruments, 15
  need for prior announcement, 16
  need to avoid redistribution, 15
Eutrophication, 8
Evolution
  economic, 91
External effects, 22

Final demand, 100
Flows, 23
Free rider, 14
Fuel efficiency, 76
Fuel mix, 76
  change, 69
Fuel mix, 76
  change, 69
  fuel requirements
    household, per unit of final demand, 149
    production, per unit of final demand, 150
Fuel use, 122
  by production sector, 122
  households, 69
  production, 69
  sectoral intensity, 123

Game theory, 22
General equilibrium models
  and aggregation, 98
  and CO_2 problems, 97
  and input-output analysis, 97
  applied (AGE), 97
Global Greenhouse Effect, 26

Subject Index

Global warming, 3
  evidence, 6
Greenhouse effect, 3
  anthropogenic, 5, 34
  global, 33
  natural, 3, 4
  role of water vapour and CO_2, 4
Greenhouse gas
  chlorofluorocarbons, 5
  CO_2, 5
  emissions, 34
  methane, 5
  nitrous oxide, 5
Gross domestic product, 48
  Germany, 64
  rate of change, 50, 54
  UK, 64

Industrial structure
Industrial structure
  change, 69
Industrialisation
  and control of nature, 7
Infrared radiation, 4

Innovation, 91, 92
  definition, 95
  of new techniques, 28
Input-output analysis, 95, 96, 99
  and activity analysis, 95
  and general equilibrium models, 97
  and national accounting, 96, 119
  direct effects, 111
  final demand, 100, 110
  indirect effects, 111
  interindustry trading, 99
  intermediate demand, 100
  Leontief inverse, 109, 110
    as infinite series, 113
  matrix form, 107
    element form, 108
    condensed form, 108
  of CO_2 emissions, 121
  price indices, 118
  prices, 113
  production assumption, 101
  production CO_2 emissions, 122
  technological coefficients, 102
    matrix, 108
Input-output data
  A matrices
    Germany, 152
    UK, 152
  B matrices
    Germany, 156
    UK, 156
  C matrices
    Germany, 153
    UK, 153
  e vector
    Germany, 155
    UK, 156
  large quantity of, 157
  m vectors
    Germany, 156
    UK, 156
  P matrices
    Germany, 154
    UK, 155
  u vectors and Y
    Germany, 156
    UK, 157
  Z matrices
    Germany, 155
    UK, 155
Input-output tables, 100
  aggregation, 151
  and value added, 116
  as simultaneous equations, 103
  balanced, 119
  deflating coefficients, 118
  from engineering data, 117
  from financial data, 117
  from value-based data, 116
  in algebraic form, 102, 103
  in value form, 114
  inter-regional trade, 133
  Leontief inverse and prices, 116
  matrix representation, 104
  output structure, 103
  two sector model, 101
Inter-industry trading, 43
Intermediate demand, 100
Intertemporal allocation theory, 22
Invention, 91
  definition, 94
  of new techniques, 28
Irreversibility, 91, 92

Lagrangian, 231
Leontief inverse, 109, 110
  as infinite series, 113, 124, 179

Subject Index

and prices, 116
Logarithmic differentiation, 49, 55, 74

Materials balance approach, 7
Matrix
  definition, 105
  inverse, 105
  inversion, 105
  multiplication, 105
  notation, 104
    condensed form, 105
    element form, 105
  singular, 106
  unit, 105
Methane, 33
  global emissions, 34, 35
Modelling
  bottom-up approach, 41
  ex ante, 26
  ex post, 26
  top-down approach, 41
Montreal protocol, 11, 16
  Montreal protocol
  and global emissions, 36

National accounting, 22, 96
  and input-output analysis, 96, 119
National income, 48, 120
  and CO_2 emissions, 47
National product, 120
Natural environment
  and production structure, 28
Natural resources, 7
Nature
  non-evaluative worth, 20
Needs
  long-run flexibility, 28
Nitrous oxide, 33
  global emissions, 35
Non-linear relationships, 24

Openness, 20
Ozone, 33
Ozone layer, 16

Pearl Harbour
  attack on, 12
Pollution
  definition of, 8
  global effects of, 9
Price indices, 118

Primal representation, 100
Prisoners' dilemma, 13
Production function, 92
  and activity analysis, 92
  linear-limitational, 25
  neoclassical, 25, 94
Production process, 25, 94
Production structure, 91
  and natural environment, 28
Property rights, 22
Proportional technology, 94, 95
Public finance, 22

Quadratic programming, 229

Scenarios
  energy efficiency, 210
    appliances, 215
    building materials, 212
    district heating, 215
    electricity, 211
    food processing, 213
    house insulation, 214
    households, 214
    industrial use, 211
    iron & steel, 212
  final demand, 205, 206
    fuels, 207
    non-fuel goods, 206
  for CO_2 emission reduction, 44
  fuel efficiency, 205
  fuel substitution, 205, 216
    biomass, 222
    geothermal, 221
    hydropower, 221
    natural gas, 217
    non-fossil fuels, 218
    nuclear energy, 219
    renewables, 220
    solar, 221
    wind, 220
  plausible, 205, 224
  private to public transport, 208
  simulations, 205
  trend extrapolations, 223
  trend forecasts, 205
Scenario analysis, 27
  minimum disruption, 227
Sea level rise, 34
Simultaneous equations
  matrix form, 106

SPIT model, 46
Statistisches Bundesamt, 65
Stocks, 23
Structural change, 12, 13, 29
  economic, 3
Synergetic effects, 44

Tableau économique, 99
Technical progress, 25
  definition, 95
Technique of production, 93, 94
Technological coefficients, 102
  imports, 134
  inter-region, 134
  intra-region, 133, 134
Technology
  back-stop, 41, 43
  change, 69
  end-of-pipe, 20, 21
  of an economy, 94
Telos, 20
Thermodynamics
  First Law of, 7, 23, 93
  Second Law of, 7, 23, 93
Toronto
  conference, 6, 11
  target, 6, 7, 11, 44, 53, 54, 207
Trace gases, 33
Tradeable emission permits, 15

Value added, 116
Values
  natural, 20
Vector
  definition, 105

Will, 11-13, 17
  impact on natural environment, 28
  meaning of, 12
Wish, 12
  meaning of, 12

List of Figures

Figure 3.1 Global CO_2 emissions by fuel 35
Figure 3.2 Global methane emissions by source 35
Figure 3.3 Global N_2O emissions 35
Figure 3.4 Global CFC(12) emissions 35
Figure 3.5 Energy use: World, USA and EC 37
Figure 3.6 Energy use (index): World, USA and EC 37
Figure 3.7 CO_2 emissions: World, USA and EC 38
Figure 3.8 CO_2 emissions (index): World, USA and EC 38
Figure 3.9 Changing energy/CO_2 ratios: World, USA and EC 38
Figure 3.10 Energy/GDP: World, USA and EC 39
Figure 3.11 CO_2/GDP: World, USA and EC 39
Figure 3.12 Changing patterns of CO_2 emissions 40

Figure 4.1 Decomposition of rate of change of World CO_2
emissions 52
Figure 4.2 Decomposition of rate of change of USA CO_2
emissions 53
Figure 4.3 Decomposition of rate of change of EC CO_2 emissions 53

Figure 5.1 Energy use by Germany and the UK 65
Figure 5.2 CO_2 emissions by Germany and the UK 65
Figure 5.3 Per capita energy use by Germany and the UK 66
Figure 5.4 Per capita CO_2 emissions by Germany and the UK 66
Figure 5.5 CO_2/energy for Germany, the UK and the EC 67
Figure 5.6 Energy use v. GDP: Germany 67
Figure 5.7 Energy use v. GDP: UK 67
Figure 5.8 CO_2 emission v. GDP: Germany 68
Figure 5.9 CO_2 emission v. GDP: UK 68
Figure 5.10 Energy use/GDP for Germany, the UK and the EC ... 68
Figure 5.11 CO_2 emission/GDP for Germany, the UK and the EC 68
Figure 5.12 Energy use by fuel type: Germany 70
Figure 5.13 Energy use by fuel type: UK 70
Figure 5.14 CO_2 emission by fuel type: Germany 70
Figure 5.15 CO_2 emission by fuel type: UK 70
Figure 5.16 Production and household energy use: Germany 72
Figure 5.17 Production and household energy use: UK 72
Figure 5.18 Production and household CO_2 emissions: Germany 72
Figure 5.19 Production and household CO_2 emissions: UK 72
Figure 5.20 Sectoral production CO_2 emissions: UK (1) 73

Figure 5.21 Sectoral production CO_2 emissions: UK (2) 73
Figure 5.22 Decomposition of total CO_2 emission changes: Germany 75
Figure 5.23 Decomposition of total CO_2 emission changes: UK ... 75
Figure 5.24 Decomposition of production CO_2 emission changes: Germany 77
Figure 5.25 Decomposition of production CO_2 emission changes: UK 77
Figure 5.26 Decomposition of household CO_2 emission changes: Germany 78
Figure 5.27: Decomposition of household CO_2 emission changes: UK 78
Figure 5.28 Decomposition of CO_2 emission changes by Agriculture: UK 79
Figure 5.29 Decomposition of CO_2 emission changes by Services: UK 79
Figure 5.30 Decomposition of CO_2 emission changes by Construction, etc: UK 80
Figure 5.31 Decomposition of CO_2 emission changes by Transport: UK 80
Figure 5.32 Decomposition of CO_2 emission changes by Paper and Printing: UK 81
Figure 5.33 Decomposition of CO_2 emission changes by Textiles: UK 81
Figure 5.34 Decomposition of CO_2 emission changes by Food: UK 81
Figure 5.35 Decomposition of CO_2 emission changes by Chemicals: UK 81
Figure 5.36 Decomposition of CO_2 emission changes by Other Industry: UK 82
Figure 5.37 Decomposition of CO_2 emission changes, including Output Mix Effect: UK 85

Figure 9.1 The shares of CO_2 emissions for Germany and the UK (1968-88) 162

Figure 10.1 CO_2 emission by Germany and the UK: domestic final demand proportion 173
Figure 10.2 CO_2 emission by Germany and the UK: domestic direct manufacturing proportion 173
Figure 10.3 CO_2 emission by Germany and the UK: domestic indirect manufacturing proportion 173
Figure 10.4 CO_2 emission by Germany and the UK: domestic total proportion 173

Figure 10.5 CO_2 emission by Germany and the UK: exports proportion 174
Figure 10.6 CO_2 responsibility by Germany and the UK: domestic proportion 176
Figure 10.7 CO_2 responsibility by Germany and the UK: direct imports proportion 176
Figure 10.8 CO_2 responsibility by Germany and the UK: indirect imports proportion 176
Figure 10.9 CO_2 responsibility by Germany and the UK: total imports proportion 176
Figure 10.10 Ratio of CO_2 responsibility to emissions: Germany and the UK 177
Figure 10.11 Contribution of changes in **C** to CO_2 emission: Germany and the UK 185
Figure 10.12 Contribution of changes in $(I - A)^{-1}$ to CO_2 emission: Germany and the UK 185
Figure 10.13 Contribution of changes in proportions of elements of **y** to CO_2 emission (**u**): Germany and the UK 186
Figure 10.14 Contribution of changes in level of **y** to CO_2 emission (Y): Germany and the UK 186
Figure 10.15 Contribution of total changes in production to CO_2 emission: Germany and the UK 186
Figure 10.16 Contribution of changes in **P** to CO_2 emission: Germany and the UK 187
Figure 10.17 Contribution of changes in **Z** to CO_2 emission: Germany and the UK 187
Figure 10.18 Contribution of changes in structure of household fuel purchases to CO_2 emission: Germany and the UK 187
Figure 10.19 Contribution of changes in level of **y** to CO_2 emission: Germany and the UK 187
Figure 10.20 Contribution of total changes in household fuel consumption to CO_2 emission: Germany and the UK 188
Figure 10.21 Total changes in CO_2 emission: Germany and UK .. 188

List of Tables

Table 3.1 Fuel units and conversion ratios to CO_2 emissions 36

Table 4.1 Decomposition of CO_2 emissions changes: World, USA and EC 54

Table 5.1 Decomposition of CO_2 emissions changes; total, production and household: Germany and the UK (1960-88) 78

Table 5.2 Decomposition of CO_2 emissions, by sector, 1970-88: UK 83

Table 5.3 Decomposition of production CO_2 emissions, by sector, 1970-88: UK 86

Table 7.1 An input-output table for a two sector model economy ... 101

Table 7.2 An input-output table for a two-sector model economy, in algebraic form 102

Table 7.3 An input-output table for a two-sector model economy, in modified algebraic form 103

Table 7.4 A value input-output table for a two-sector model economy, in algebraic form 114

Table 7.5 An illustrative value input-output table for a two-sector model economy 119

Table 8.1 An inter-regional trade input-output table 133

Table 9.1 Value of output and fuel use: Germany 1988 159
Table 9.2 CO_2 intensities: Germany 1988 160
Table 9.3 CO_2 emissions: Germany 1988 163
Table 9.4 Ranked CO_2 intensities: Germany 1988 165
Table 9.5 Ranked $CO2$ emissions: Germany 1988 166
Table 9.6 CO_2 emissions: elasticities with respect to final demand, fuels, energy, column sums and row sums: Germany 1988 168

Table 10.1 Emissions of CO_2 to satisfy domestic and export final demand: Germany and the UK 172

Table 10.2 Responsibilities for CO_2 Emissions: Germany and the UK 175

Table 10.3 Attribution of difference in CO_2 emission between Germany 1986 and Germany 1988 181

Table 10.4 Attribution of difference in CO_2 emission between UK 1979 and UK 1984 182
Table 10.5 CO_2 emission changes: Germany and the UK 184
Table 10.6 Attribution of difference in CO_2 emission between UK 1979 and Germany 1980 190
Table 10.7 Attribution of difference in CO_2 emission between UK 1984 and Germany 1984 191

Table A10.1 Attribution of difference in CO_2 emission between Germany 1978 and Germany 1980 194
Table A10.2 Attribution of difference in CO_2 emission between Germany 1980 and Germany 1982 195
Table A10.3 Attribution of difference in CO_2 emission between Germany 1982 and Germany 1984 196
Table A10.4 Attribution of difference in CO_2 emission between Germany 1984 and Germany 1986 197
Table A10.5 Attribution of difference in CO_2 emission between Germany 1986 and Germany 1988 198
Table A10.6 Attribution of difference in CO_2 emission between UK 1968 and UK 1974 199
Table A10.7 Attribution of difference in CO_2 emission between UK 1974 and UK 1979 200
Table A10.8 Attribution of difference in CO_2 emission between UK 1979 and UK 1984 201

Table 11.1 Change in CO_2 emissions with tertiary sectors expansion, given GDP growth rate of 2% p.a 207
Table 11.2 Possibilities for CO_2 emission reduction by altering the structure of final demand for fuels, for a given level of GDP .. 208
Table 11.3 CO_2 emission reductions due to altering the structure of final demand: private to public transport 209
Table 11.4 CO_2 emission reductions by changing from average to best practice technology: Changes in C 214
Table 11.5 CO_2 emission reductions by changes from average to best practice technology: Changes in P 216
Table 11.6 CO_2 emission reductions by changing the energy mix: Gas for coal in industry 217
Table 11.7 CO_2 emission reductions by changing fuel mix: Substitution of fossil fuels by gas in industry and final demand 218
Table 11.8 CO_2 emission changes by substituting fossil fuels for nuclear fuels in energy generation 219
Table 11.9 CO_2 emission reductions by switching the energy mix to non-fossil fuels 222

Table 11.10 Sectoral growth rate assumptions for the 'plausible scenarios' 224
Table 11.11 A sequence of 'plausible' scenarios for CO_2 emission reductions in Germany and the UK 225

Table 13.1 Minimum disruption changes to final demand: 1% reduction in CO_2 emission; Germany and the UK 249
Table 13.2 Minimum disruption changes to final demand: 1% reduction in CO_2 emission by Germany and the UK; subject to 2% growth in GDP 251
Table 13.3 Minimum disruption changes to final demand: 1% reduction in CO_2 emission by Germany and the UK; subject to no change in employment. 253
Table 13.4 Minimum disruption changes to final demand: 1% reduction in CO_2 emission by Germany and the UK; subject to 2% growth in GDP and employment 254
Table 13.5 Actual annualised rates of change of final demand: Germany (1978-88) and UK (1968-84) 256
Table 13.6 Minimum disruption changes to fuel use (i.e C): 1% reduction in CO_2 emission by Germany 258
Table 13.7 Minimum disruption changes to fuel use (i.e C): 1% reduction in CO_2 emission by the UK 259
Table 13.8 Minimum disruption changes to fuel use (i.e C): 1% reduction in CO_2 emission by Germany; subject to no change in overall energy efficiency by each sector 260
Table 13.9 Minimum disruption changes to fuel use (i.e C): 1% reduction in CO_2 emission by the UK; subject to no change in overall energy efficiency by each sector 261
Table 13.10 Actual changes to fuel use (i.e C): Germany 262
Table 13.11 Actual changes to fuel use (i.e C): UK. 263
Table 13.12 Minimum disruption changes to interindustry trading (i.e. A): 1% reduction in CO_2 emission by Germany and the UK ... 265
Table 13.13 Actual annualised rates of change of the row and column sums of the A matrices: Germany (1978-88) and the UK (1968-84) 266